Astronomers' Universe

Series Editor
Martin Beech, Campion College
The University of Regina
Regina, Canada

The Astronomers' Universe series attracts scientifically curious readers with a passion for astronomy and its related fields. In this series, you will venture beyond the basics to gain a deeper understanding of the cosmos—all from the comfort of your chair.

Our books cover any and all topics related to the scientific study of the Universe and our place in it, exploring discoveries and theories in areas ranging from cosmology and astrophysics to planetary science and astrobiology.

This series bridges the gap between very basic popular science books and higher-level textbooks, providing rigorous, yet digestible forays for the intrepid lay reader. It goes beyond a beginner's level, introducing you to more complex concepts that will expand your knowledge of the cosmos. The books are written in a didactic and descriptive style, including basic mathematics where necessary.

Günther Rüdiger

The Astrophysical Observatory Potsdam - Triumph and Tragedies

Reports and Memories 1874 - 1991

Günther Rüdiger
MHD and Turbulence (ret.)
Leibniz-Institut für Astrophysik Potsdam
Potsdam, Germany

Based on a translation from the German language edition: "Astronomen, Akten und Affären" by Günther Rüdiger, © AVA Akademische Verlagsanstalt 2024. Published by AVA Akademische Verlagsanstalt 2024. All Rights Reserved.

ISSN 1614-659X   ISSN 2197-6651   (electronic)
Astronomers' Universe
ISBN 978-3-032-06293-2   ISBN 978-3-032-06294-9   (eBook)
https://doi.org/10.1007/978-3-032-06294-9

The Astrophysical Observatory Potsdam in the first half of the 20th century with Great Refractor in the background. Courtesy of H. Sperlich née Starke

This Springer imprint is published by the registered company Springer Nature Switzerland AG
The registered company address is: Gewerbestrasse 11, 6330 Cham, Switzerland

If disposing of this product, please recycle the paper.

The Astrophysical Observatory Potsdam in the first half of the twentieth century with Great Refractor in the background. Courtesy of H. Sperlich née Starke

For Gisela, Anna, Luisa and Moritz

Preface

As is common among astronomers, the author of this book has spent his life working at various aptly named institutes such as the Astrophysical Observatory, the Institute for Stellar Physics, or the Central Institute for Astrophysics. Yet, it was always in Potsdam and always the same institution— only its directors and names changed over time. The observatories on the Telegraphenberg in Potsdam and on Erbisbühl in Sonneberg, which once briefly formed the Institute for Stellar Physics of the Academy of Sciences have since vanished as institutions; only their domes and buildings remain. One might be tempted to recount modern German history through the CVs of its astronomers and their observatories to show how the intellectual level of a nation is reflected by its scientific structures.

This is precisely what has been attempted here, by retelling the triumph and tragedies of the astronomers at the Astrophysical Observatory at Potsdam (AOP). Founded almost exactly 150 years ago as a compromise between two opposing visions, a never-ending dispute began when, in early 1882, one of the originally equal Observers was appointed sole director. Despite this, a very small group of researchers revolutionized solar and stellar physics with extraordinary diligence: they discovered extremely close binary stars through spectral analysis, identified interstellar matter, and recognized the decades-long pause in solar activity now known as the Maunder Minimum. The frequently public quarrels reached a dramatic climax when Erwin Freundlich, head of the Einstein Tower, insisted on his institute's independence supported by the state science administration. The director of the Observatory, Ludendorff, opposed this unsuccessfully until the times turned in his favor around 1933.

The idea that many properties of the Sun and stars are fundamentally shaped by magnetic fields gained wide acceptance only a decade later at the AOP. Harald von Klüber installed polarization optics in the telescope of the Einstein tower 35 years after sunspot magnetism had been discovered at Mt. Wilson, thereby initiating the study of cosmic magnetic fields in Potsdam—the eventual "DNA" of the late AOP. "The magnetohydrodynamics of turbulent flows, which is particularly important in astrophysics, is still a closed book to us," lamented Albrecht Unsöld in 1957. Just 3 years later, Max Steenbeck in Jena eagerly took up this challenge at his Institute for Magnetohydrodynamics of the Academy of Sciences at Berlin. His successors would later arrive at an almost empty Observatory in Potsdam in an effort to restore its international visibility by producing new insights into the magnetic field generation of planets, stars, and galaxies via the dynamo effect.

Shortly before the Berlin Wall was erected in 1961, many leading scientists had left the Observatory for West Germany. The Wall would fall only a generation later, shaped in part by the spontaneous emergence of new political movements in the GDR. Two of the early activists had come from the AOP itself, prompting the institute's chief ideologue to complain: "Four of the so-called signatories are employed at our Academy—two of them in astrophysics alone. What a shame for our institute."

East German astronomers were under close surveillance by the GDR's state security forces, particularly after the IAU General Assembly in Prague in 1967. The agents rightly suspected that hardly any GDR astronomer believed their work could have long-term meaning without contact with the Western scientific community. It may have been the first—and perhaps the last—time that in 1989 (middle-aged) astronomers descended from their mountain-top observatories to play a leading role in reshaping the political order of their country and, in doing so, took control of their own future.

The final part of the book recounts how, despite these developments, the internationally renowned AOP eventually disappeared from the scientific landscape altogether, ceasing to exist by the end of 1991. In the world of collectors, the AOP has since become a closed collecting field. This text is thus also an attempt to mentally reconnect the buildings and facilities on Potsdam's Telegraphenberg with their scientific heritage especially since, to this day, it has not been possible to honor the legacy of pioneers such as Vogel, Spörer, Hartmann, Schwarzschild, Hertzsprung, Freundlich, von Klüber, Grotrian, and Krause with anything more than written tributes.

Potsdam, Germany   Günther Rüdiger
17 June 2025

Competing Interests The author has no competing interests to declare that are relevant to the content of this manuscript.

Contents

Abbreviations

ABBAW	Archive of the Berlin-Brandenburg Academy of Sciences
ADS	Astrophysics Data System
AdW	Academy of Sciences of the GDR
AG	German Astronomische Gesellschaft
AIP	Astrophysical Institute Potsdam
AOP	Astrophysical Observatory Potsdam
AVUS	Automobile traffic and practice road (motorway in Berlin)
BAT	Bundes-Angestelltentarifvertrag (collective pay agreement)
BL	Blaue Liste (List of German institutions first written on blue paper)
BLHA	Brandenburg State Main Archive
BMFT	Federal Ministry of Research and Technology
BStU	Stasi Records Archive (formerly: Office of the Federal Commissioner for the Records of the MfS of the former GDR)
BV	Bezirksverwaltung (regional administration of the MfS)
CCD	Charge-coupled Device
CdC	Carte du Ciel
CDU	Christian Democratic Union of Germany
COSPAR	Committee on Space Research
ČSSR	Czechoslovak Socialist Republic
DARA	German Agency for Space Affairs
DAW	German Academy of Sciences
DEFA	Deutsche Film AG
DFA	pressing family matters
DFG	German Research Foundation
DLR	German Aerospace Center
ESA	European Space Agency
ESER	Unified System of Electronic Computing Technology Company
FDJ	Free German Youth

FIM	Hidden Principal Informer
FRG	Federal Republic of Germany
FSU	Friedrich Schiller University Jena
GAFD	Geophysical and Astrophysical Fluid Dynamics
GDR	German Democratic Republic (DDR)
GS	General Secretary
GStAPK	Geheimes Staatsarchiv Preußischer Kulturbesitz
HAO	High Altitude Observatory
HHI	Heinrich Hertz Institute
HMI	Hahn-Meitner Institute
IAU	International Astronomical Union
IKF	Institute for Cosmos Research
IM	Unofficial employee of the MfS
IPCC	Intergovernmental Panel on Climate Change
ISC	Internal Scientific council of ZIAP (1990/1991)
MfS	Ministry for State Security of the GDR
MHD	Magnetohydrodynamics
MPE	Max-Planck-Institute for Extraterrestrial Physics
MPG	Max Planck Society (Max-Planck-Gesellschaft)
MWFK	Ministry of Science, Research and Culture
NCAR	National Center for Atmospheric Research
NKA	National Committee for Astronomy
NS	National Socialist
NSDAP	National Socialist German Labour Party
OAO	Objedinjonnaja Astronomitscheskaja Observatorija (United Observatory of Socialist Countries)
OV	Hidden Operation of the MfS
OPK	Operational procedure of the MfS
RDS	Rat Deutscher Sternwarten
RM	Reichsmark
ROS	Council of East German Observatories
ROSAT	ROentgenSATellit
SA	Sturmabteilung (NSDAP)
SDP	Social Democratic Party in the GDR (1990)
SED	Socialist Unity Party of Germany
STELLA	STELLar Activity
Thlr	Thaler
USSR	Union of Soviet Socialist Republics
UV	Ultraviolet
VAX	Virtual Address eXtension (computer architecture)
VEB	Volkseigener Betrieb
VIP	Very Important Person
WIP	Scientist Integration Programme

WR	German Science and Humanities Council (Wissenschaftsrat)
ZIAP	Central Institute for Astrophysics
ZIPE	Central Institute for Physics of the Earth
ZISTP	Central Institute for Solar-Terrestrial Physics

List of Figures

Part I

How Astrophysics Came to Potsdam

1

Prelude: What Is a Sun?

The binary star researcher William Herschel from Slough was also an enthusiastic sunspot observer. In 1801, 20 years after his discovery of the planet Uranus, he presented careful observations of the sun's surface in a detailed study because "it becomes almost a duty for us to study … the solar surface."[1] With precise drawings of the "openings," Herschel documented his particular interest in the occurrence of sunspots, particularly over longer periods of time, convinced of their influence on global weather and food production (Fig. 1.1). Due to a lack of climate data, he used the grain prices compiled by Adam Smith and, as suspected, found high prices when the sun was spotless—and the sun was spotless more often, at least in the past: The historical material contains 5 irregularly spaced and unequally long periods without sunspots. The first lasted from 1650 to 1670. The next … lasts until 1684 … One spot in 1710, none in 1711 and 1712 and another in 1713 … Remarkably, after this time there were never again no spots. Herschel would certainly have found the 11-year sunspot cycle if the sun had not happened to be in a grand minimum of its magnetic activity during his inspections. To him, the spots were randomly appearing openings in the sun's atmosphere—revealing the dark solar body below—and a periodic variation in their occurrence would have seemed unreasonable or unexplainable. After all, he had been observing the surface of the sun for 40 years, just as Samuel Heinrich Schwabe did later.[2]

In a subsequent "Lecture to a mixed Audience," R. Wolf found Herschel's opening model appealing, especially as the dark core of the spot, the umbra,

---

[1] Herschel (1801).

[2] For details of data collection and of the solar observers see Arlt & Vaquero (2020).

Fig. 1.1 Left to right: William Herschel (1738–1822); Herschel's 10-foot telescope (22 cm metal mirror). The sunspot observations published in 1801 were made with such an instrument. (Photo Universitätssternwarte Göttingen)

really did appear to lie below the surface of the sun, but he destroyed with statistics the idea of the existence of a correlation between spot frequency and weather. Around 1861, his compilation of sunspots included observations "for 2143 days from the 17th century, 5490 days from the 18th century and for 14,860 from the current century." Wolf must have been very fair and the postal deliveries must have been very fast and reliable. Just a few months after its publication, he circulated the latest spectacular results from the private astronomer Richard Carrington in Redhill near London (Fig. 1.2). In a first, since 1853 Carrington had determined and calculated thousands of positions and proper motions of the spots in longitude and latitude with his 12-cm-aperture telescope, for which he had to determine the position of the sun's axis of rotation. He found the differential solar rotation: his plots showed that the daily rotational motion clearly depended on the solar latitude. Before his publications "there was still not the smallest suspicion that the differences of period observed ... depended in any way on the latitude of the particular spot."[3] He also realised that sunspots were never visible at the equator or the poles. They only appeared in belts between 6° and 35° north and south latitude, which themselves slowly moved towards the equator and eventually disappeared there. "Whether this is what occurs at each period of increase and decrease of the frequency of the spots must be left to observers who may follow me to show. At present it is only probable that such is the case, and

---

[3] Carrington (1863).

Fig. 1.2 Redhill observatory of Richard Christopher Carrington (1826–1875), of whom neither a photograph nor a portrait seems to exist (Cliver, 2005). Even the obligatory painting from the Royal Society's collection is said to have disappeared at some point

another contribution made to the facts … will ultimately elucidate the origin of this phenomenon and instruct us on the question, 'What is a Sun?'."

Carrington was highly interested in the sun's variable surface: The observation of the stars requires the hours of the night and afforded little matter of speculation; the observation of the Sun was a day-task, and presented more variety and interest.[4] He made two fundamental discoveries in the early solar physics in only 7 years of work: the nonuniform ("differential") rotation of the solar surface and the latitudinal migration of the spot zones. The fact that the sun does not rotate like a rigid body but that the equator regularly overtakes the polar regions, is said to have changed the image of the sun more drastically than the discovery of the Fraunhofer lines.[5] A giant work, as if it had been too much: 4 years later, he fell ill and died at the age of 49, just a few days after his wife and in the same year as the elder Samuel Heinrich Schwabe. The Royal Society had awarded the latter its golden medal, which has been brought to Dessau by Carrington in 1857. Despite his immense successes, he

---

[4] See "Richard Christopher Carrington (obituary)", Monthly Notices of the Roy. Astron. Soc. 36, 137 (1876).

[5] Clerke (1887).

was never granted a position as a professional astronomer—even though there were vacancies in Oxford in 1859 and Cambridge in 1861.[6]

The unhappy pharmacist Samuel Heinrich Schwabe in Dessau had already achieved his masterpiece in the 1840s (Fig. 1.3). Schwabe, during his school days, assisted his father with operations and to glue bags for his grandfather. Remarking about his faded colleagues, Wolf explained: "While Schwabe's predecessors either neglected to draw a complete picture of the sun day by day, and mostly proceeded from the opinion that it was only interesting if the sun showed many spots, he did not let a day go by for 35 years[7] without making sure whether the sun showed spots or not."[8] Schwabe's preference for all sorts of statistics had revealed a well-hidden law from the heaven which many observers could have found.

At the beginning of his records, in 1828, Schwabe had seen sunspots on every observation day, then less and less frequently. Five years later he observed them only every second day, but in 1837 again they appeared daily. After another 5 years, in 1843, there were only a few again. He always wrote everything down in extensive annual diaries, even if he saw nothing. Should a particularly large number of spots cover the sun every 10 years? He had only really seen one single maximum, that of 1837. He appeared to have missed the 1826 maximum, however Schwabe was careful, he couldn't be wrong, otherwise the palace administration, from whose income he lived, might

Fig. 1.3 Left to right: Samuel Heinrich Schwabe (1789–1875); Dessau, Johannisstraße, roof astronomy with dramatic results

---

[6] Cliver & Keer (2012).

[7] First observation on 5 November 1825.

[8] Wolf (1861).

abandon his salary. On the other hand, it would be a huge task to find a real law of nature, it was too tempting and as a trial he wrote that the sunspots have a period of about 10 years. His talents had always been his sense of order; would anyone believe that he had found a clock in the sun? Just a few sentences for the Astronomische Nachrichten in Hamburg-Altona: If one now compares the number of groups and the number of spot-free days, one finds that the sunspots had a period of about 10 years ... The future must teach us whether this period shows some consistency.[9] And to Humboldt, cautiously he wrote that he had not an opportunity to become acquainted with older observations in a continuous series, "but I readily agree with the opinion that this period itself may again be variable."[10] While the original communication in the Astronomische Nachrichten was mainly ignored,[11] Humboldt's assistance—he presented Schwabe's numbers in his "Cosmos" in 1850—brought the desired public acceptance. It was at the end of 1868, when he was no longer able to climb his rooftop observatory, that he had to cease his observations, after more than 40 years of "Keplerian faith in the laws and order of Nature."[12] Schwabe donated his instruments to the gymnasium; he died in 1875 at the age of 86 (Fig. 1.4).

---

[9] Schwabe (1844).
[10] Humboldt (2004), volume III, p. 543.
[11] The first citation of this paper in the Astrophysics Data System (ADS) dates from 1948.
[12] Clerke (1887).

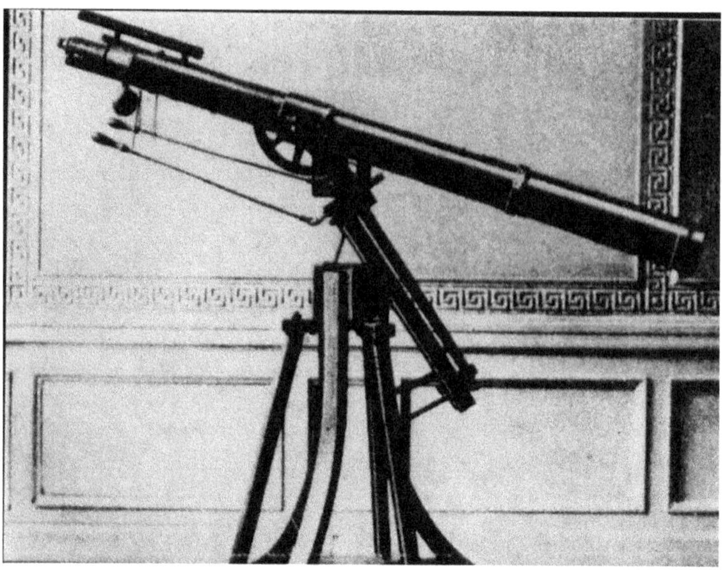

Fig. 1.4 The larger of Schwabe's two instruments (amplification 64) from Dresden. Schwabe used filters from Fraunhofer. (Archive Erhard Hirsch)

2

From the Sunspots to the Stars

Escape North

In the fall of 1864 during the German-Danish war a 16-year-old boy received a provisional certificate from the reform grammar school in Anklam because his widowed mother could no longer pay the school fees. The certificate was mediocre: from maths he had geometry up to the theory of similarity, arithmetic up to simple equations, but for drawing he always showed a lot of participation and skill. Later, the school would be proud to carry the name of this student, and a personal museum in Anklam will exhibit glorious proofs of his genius. The student, Otto Lilienthal, travelled to Potsdam to a much cheaper, more craft-oriented school which he graduated from after 2 years, before returning home to start flight experiments that would later, in his 48th year, costs him his life.

The Hanseatic town Anklam had only been connected to Berlin and Potsdam by the Stralsund railway the previous year. The King came from Potsdam to the inauguration of the new railway station in March 1863, the whole town was on its feet, the younger pupils sang and the grammar school students threw their caps in the air, and the largest ship ever built in Anklam was launched in the shipyard. Lilienthal's maths teacher Dr. Spörer[1] from Berlin, who did not always understand the idiom of his student, had been awarded the title of professor in the same year, 15 years after he had become

---

[1] In numerous documents from the nineteenth century, including private ones, the spelling is "Spoerer" instead of the "Spörer" used today. Similar Foerster → Förster.

G. Rüdiger, *The Astrophysical Observatory Potsdam - Triumph and Tragedies,* Astronomers' Universe, https://doi.org/10.1007/978-3-032-06294-9_2

a substitute teacher at this gymnasium at 1849. It was only the previous year that the school, which had started teaching with 5 particularly ambitious teachers and 300 students, had been granted the full rights for high-schools. Spörer had received his permanent position at the school in November 1853 as a widower (Fig. 2.1). At the age of 30, he had married a daughter of a medical officer from Magdeburg, Louise Auguste, who died immediately after the birth of her daughter Marina—his first marriage only lasted the 9 months of pregnancy.

In addition to the 13 lessons a week for maths and 6 lessons for physics, Spörer still found time to write a treatise on his teaching concept, which he presented to the school management in January 1855. "By introducing the student to maths, he is ruled to move in abstractions." This was followed by a detailed description of his personal curriculum: I started with [differential calculus] once in the primary, but after only six lessons I was fully convinced that too many were unable to follow. The diplomatically formulated letter seemed to have been liked in Anklam and Berlin. In his 34th year, students and parents were informed that on 31 May 1855 the mathematician Dr. Spörer in recognition of the success of his activities received by the Minister v. Raumer the title of a schoolmaster.

"Tests that I have to correct at home will only be delivered from secondary school onwards" is announced in his instructions for maths teachers. With almost no class work to correct and only 19 lessons a week on 6 school days, he must not have been working too hard. Without differential and integral calculus, all that remained was geometry, simple algebra and trigonometry, all of which he had learnt from Prof. Encke at the observatory in Berlin as the

Fig. 2.1 Gustav Spörer as a teacher; gymnasium in Anklam, built in 1850/1851. (Courtesy of W. Hornburg)

necessary tools of the astronomers. He certainly did not have to prepare for his lessons. After 3 years of study, he graduated in 1843 at the age of just 21 with a numerical dissertation on a comet's orbit. Until the fall of 1845, Spörer was assisting at the Sternwarte Berlin, but as there was no hope of being employed in Berlin or at another observatory, "he took the examination pro facultate docendi,"[2] the teaching licence. The disappointment of not being able to work as an astronomer must have shaken the high-talented graduate to the core; he decided to get away from universities and academies, at least far away from Berlin. In order to find a job as a gymnasium teacher, he turned northwards—maths teachers were always in demand—and settled in Bromberg and Prenzlau, but in 1849 he found a job even further north, initially a temporary one in Anklam. This escape, however, will bring him back to Berlin decades later with his then extended family in triumph.

Even during Spörer's first student year, K.H. Schellbach became a professor of mathematics at his former Gymnasium in Berlin. Schellbach founded—if he had read Spörer's tractate—a (short-lived) pedagogical seminar in 1855 to train young mathematicians for school teaching in order to raise mathematical education to the level of training the old languages. Higher ambitions in education, such as those of the reforms in Thuringia, were still a long way off in Prussia. Schellbach later taught the prince Frederick[3] privately in the natural sciences, often in the presence of the interested mother. Astronomy was also part of the programme since the discovery of spectral analysis by Kirchhoff and Bunsen in 1861 "the study of the sun's surface has provided more and more results. Professor Spörer in Anklam was mainly concerned with the study of sunspots and prominences. But his optical apparatus was not powerful enough to work quickly and successfully." The crown prince decided to give Professor Spörer a larger telescope.[4] Already in 1863 there had been activities to give him a *7-foot Fraunhofer telescope* from the *physics collection of the local university.* The administration, however, had a long list of its timeless concerns against this plan: Does a teacher have enough time to use the telescope regularly? Is there a suitable building in Anklam with a diameter of at least 12 feet? Who will pay for the metal tripod? The plan failed because of the toxic argument, that somebody *who has so much scientific success will hardly stay in Anklam for long, as it must be his ambition to move to a larger town.*[5]

---

[2] Lohse (1895).

[3] The later Kaiser Frederick III.

[4] Schellbach (1890). The donation consisted of the balance of the costs.

[5] GStAPK File I. HA Rep. 76, Va Sekt. 2 Tit. VII, No. 14, Vol. 2. Texts from archival documents are italicised throughout.

Sunny Anklam

The gymnasium—probably at Spörer's suggestion—had already applied for a grant from the Anklam town for a high-quality school telescope from Steinheil in 1860 and received the full amount, so that a tripod could be financed separately.[6] It was a 3½-foot refractor[7] with a circular micrometer, just right for determining the exact location of sunspots. Spörer as a teacher only had time in the afternoons, and he had heard about the Schwabe cycle and also about Carrington's existential problems—he had to stop working and sell the observatory. Solar observations were thus the only realistic way of staying in astronomy; it was also obvious to perfect the measurements of the sunspot coordinates in order to determine the rotation of the sun, Carrington's discovery[8] of the migration of the two spot zones to the equator and to study the still unknown proper motions of the sunspots. He wanted to do this after class, when the majority of his fellow teachers had gone to bed. All he had to do was to cross the empty market square, a hectare in size, to the old Anklam powder tower, which had been given a new staircase and on which a glazed round building with an easy-to-open south-west side had been built around a massive tripod (Fig. 2.2). This was his observation station for a decade and a half, which had to withstand many storms. The telescope only began to shake in strong southerly winds, while the tower walls were thick enough to absorb the vibrations caused by road traffic. The city, *guided by its interest in science*, had the half-ruined tower carefully restored *for unlimited use* at its own expense. The observation pavilion, which was firmly connected to the new ceiling construction, had been built at ground level as an enclosure for the school telescope. Soon it became obvious that Anklam, with the nearby island Usedom, was one of the sunniest areas of Germany and thus quite suitable for solar studies.

Spörer's observations in December 1860 coincided with a spot maximum, so that the rich yield of the observation campaign must have facilitated the support of the school management and the help of students. He published very detailed observation results in the annual programmes of his gymnasium—certainly not against the will of his rector. At the beginning there were many spots to be seen, almost all of which appeared about 16° north or south of the equator. Five years later, there were only a few spots left, and they all

---

[6]Gymnasium zu Anclam 1861 "Invitation to the public examination of all classes," W. Dietze Anclam.

[7]More precisely: 42″ = 106 cm focal length, just under 3″ = 7.6 cm aperture (33 lines). A letter reports that he also privately owned a 2½-foot refractor.

[8]Carrington (1859).

Fig. 2.2   Left to right: Powder tower in Anklam with self-made semi-professional solar observatory; publication of the Astronomische Gesellschaft. (Courtesy of W. Hornburg)

appeared very close to the equator. Carrington's law of equatorial spot zone migration was thus confirmed for the southern and northern solar hemispheres of the new solar cycle.[9] Both observers had noticed almost simultaneously that the spots near the equator rotate faster than those further away from the equator. Just 6 months after beginning of his campaign, Spörer wrote: Of the spots listed above, those belonging to an equatorial zone therefore have a smaller rotation time than the others[10] and this is also what it says in his very first astronomical publication. A furious start of the new part-time astronomer, who was 40 years old and found the fundamental law of star surfaces to rotate differently from a rigid body. He also tried out the simplest trigonometric expressions for the newly discovered rotation law, but could not decide between $\cos(b)$ or $\cos(2b)$ for the angular velocity of rotation.[11] Here b is the heliographic latitude. The difference can hardly be seen in the data from 1860 to 1866 due to the proximity of the spots to the equator. Later he stated that he had seen the maximum latitude of a spot as 40° in

---

[9] Often called Spörer's law, today butterfly diagram.

[10] Spörer (1861).

[11] Kempf (1916).

1869. The data provided by Carrington and Spörer from the spot's proper motion are almost identical, although they were derived in different decades. He described equatorial westerly storms and easterly storms at higher latitudes, which may drive the cloudy patches, and also their individual latitudinal and longitudinal motions did not remain hidden from him. It was probably because of these random velocities that he had not been able to explicitly write down the sun's law of rotation in its later valid form, despite his enormous archive of data.[12]

Schellbach continued to inform the prince about Spörer's discoveries. Whether the information from Anklam to Potsdam went directly or via Wilhelm Förster[13] has remained unknown. In a surviving draft of a letter dated 15 December 1863 to a *High Ministry*, Spörer complained that a *telescope such as I now have at my disposal, namely a 3½-foot telescope belonging to the gymnasium, is far from sufficient for the faint edge of the sun*. With regard to sunspots, *it should* be *noted that their observation and description* have *been undertaken many times, but always too little* has *been measured*. The most careful measurements would be necessary in order to be able to judge the nature of the sun. The advertising was successful: in the second paragraph of the summary of the Anklam results[14] it is said that in 1865 he received a 7-foot telescope,[15] with which now the locations of the more important spots could be measured with high accuracy (Fig. 2.3). The prince had paid for an

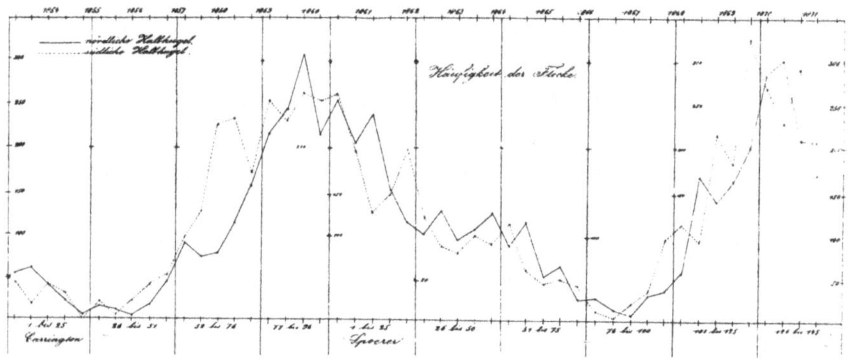

Fig. 2.3 Anklam sunspot statistics for 1854–1871 with the northern and southern hemispheres separated (Spörer, 1874)

---

[12] R. Carrington preferred the expression $\cos^{7/4} b$ for the variation of the solar rotation with latitude b in a short communication in 1862 (MNRAS 22, 300).

[13] Herrmann (1975).

[14] Spörer (1874).

[15] Focal length 213 cm, one often finds the wrong year 1868 for the donation.

instrument from the famous factory of Steinheil & Sons in Schwabing. Spörer will transport it to India for the solar eclipse in the summer of 1868 and later took it to Potsdam, where it would remain in service for decades.[16] Because of the solar eclipse expedition, it received a new tripod with clockwork at royal expense.

Rainfall Eclipse

Immediately after receiving the new telescope, Spörer wrote a letter to the director of the Sternwarte Berlin, Wilhelm Förster, the successor of his doctoral supervisor Encke. Förster was then the secretary of the Astronomische Gesellschaft founded 1863 in Heidelberg on his initiative and that of his former student Bruhns, now in Leipzig, in accordance with Saxon association law. Twenty-six astronomers had gathered in Heidelberg for the founding, and the membership grew to 149 men in the same year,[17] including Spörer who, however, was only rarely able to attend external meetings due to teaching commitments. At the first meeting in Leipzig at the end of August 1865, Förster informed the audience of a letter addressed to him from the member Prof. Spörer in Anklam.[18] The letter concerned a proposal following from his sunspot observations to set up a special institute to study the nature of the sun. Prof. Spörer asked the society to declare itself in favour of the necessity of such a solar observatory, where photometric and spectral observations should be systematically carried out in addition to measurements and recordings of the spots.[19] The assembly agreed with the wish that Prof. Spoerer may soon be put in a position to continue his work under even more favourable conditions. Not a very promising result, except that from now on his name was known to all German astronomers.

Nevertheless, an angel had guided Spörer's hand. In May 1868, following a petition from a member of the Astronomische Gesellschaft to the Reichstag of the newly founded, short-lived and Prussian-dominated North German Confederation based in Berlin, the Reichstag decided to finance a North German astronomical expedition to distant regions to view the total solar eclipse on 18 August of the same year (!). The professor from Anklam was to lead the eight participants nominated by the board, half of whom were to

---

[16] In the east dome of the main building of the AOP, together with a large spectroscope for prominence observations, from 1908, the so-called Zeiss-triplet (15 cm refractor) for photography/photometry.

[17] Constituent members, see Schielicke (2013).

[18] The original letter could not be found to date.

[19] Vierteljahrsschrift d. Astron. Ges. 1, 4 (1866).

travel to Aden in Yemen and half to Mulvad in the Middle East due to the weather risks. The majority of the teams were from Berlin, supplemented by one participant each from Anklam, Leipzig and Bonn. The Reichstag had commissioned W. Förster to submit a concept and a cost estimate. It would be necessary to "set up the finest measuring equipment for spectral analysis and the use of photography" and this equipment would have to be purchased. He claimed costs for the photographic apparatus, for the spectral apparatus and for the new mounting of the existing telescopes to 5200 Thlr, 9700 Thlr were the travel expenses and 1000 Thlr needed for the preparations. The chancellor of the North German Confederation immediately replied: I am pleased to inform the board of the Astronomische Gesellschaft … that the governments of the North German Confederation have decided to grant the project … in the total amount of 16,000 Thlr.[20]

However, there were only 4 weeks left until departure, and some board members refused to support this adventure. In fact, the spectral equipment had arrived too late, so that the necessary adjustment of the devices and training of the observers was hardly possible. The expected eclipse times at the expeditions' destinations were only 3 or 5 min minus the excitement of all participants, which, as experience had shown, always occurs.

Spörer's absence from his gymnasium because of leading a scientific expedition lasted from 4 July to 12 October 1868. The journey began on the evening of 8 July in Berlin by train to Trieste, from Trieste by ship to Alexandria, then by train with a stop in Cairo to Suez—the Suez Canal was not opened until the following year—and by steamship to Mumbai, arriving on 30 July. The expedition reached its destination by train, camel and ox cart more than 4 weeks after the start of the journey.[21] The jungle tour with heavy luggage was an organisational challenge that could only be overcome with the help of numerous local staff. It permanently rained until the day before the expected eclipse, but the pillars still had to be built, the observation huts for the instruments erected and the tents set up for the people. The morning of the eclipse day was initially cloud-free, but 10 min before first contact the sky closed in again and at 7.50 a.m. the eclipse began unnoticed. But in a gap in the clouds, the corona suddenly appeared in the sky, "surrounding the lunar disc as a dimly shining radiant wreath." Spörer saw a large eastern prominence through his eyepiece and, as his personal main result, could measure its height; his colleague Tietjen saw prominences on both the eastern and western edges of the sun; spectral images were completely omitted in the rush. It was now certain

---

[20] Vierteljahrsschrift d. Astron. Ges. 3, 186 (1868); one of the first grants for a scientific project in Prussia.
[21] Spörer (1869).

that the prominences belonged to the sun and were not phenomena of the earth's atmosphere. Not a single photo from this expedition exists; all expedition photographers had been sent to the parallel event in Aden.

Other observers had more success. A 1000 km from the German base camp, on the east coast of India, the French expedition had caused the sensation of the 1868 eclipse campaign. Their spectral apparatus showed a yellow line near the wavelength of sodium for the prominence at the edge of the sun, which was more extended the more radially the slit was orientated. In contrast, the line was only very short when the slit was directed tangentially to the solar limb; the prominence apparently had the shape of long radial fingers of unknown temperature. Jules Janssen repeated the observation the next day and saw the prominence again in the light of the yellow spectral line, even without eclipse. The line, as it later turned out, belonged to an element still unknown on Earth and was named helium.[22] Hence, the total solar eclipse of 18 August 1868 produced far more important results than any of the former expeditions.[23] The German Astronomische Gesellschaft summarised that the overall success of the North German expeditions cannot be considered satisfactory. Nevertheless, the spectral apparatuses brought along were such that under more favourable weather conditions the important discovery of the typical character of the prominence spectra "could not have escaped our observers."

After his return to Anklam on 12 October, colleagues and students greeted him with joy; in his school speech the previous year, the director only had to remember the four former pupils felt in a war battle.[24] At the end of the 1868/69 school year, Spörer was elected to the vacant position of the prorector. From then on, he had to accompany the students of the upper classes on their annual trips to German islands in the Baltic Sea, probably secretly hoping for rain so that he would not lose any worth observation time.

What is Astrophysics?

In the course of formation of the Deutsches Reich at the beginning of 1871 almost 2 billion marks had flowed to the former North German Confederation. Prussia's Wilhelminian time began and prince Friedrich could start to realise his cultural and scientific projects. At Schloss Babelsberg, people remembered the restless solar observer in Anklam, his unhappy Indian expedition, but

---

[22] N. Lockyer found this spectral line 1868 even without an eclipse and suggested the name helium.

[23] Klein (1869).

[24] Gymnasium zu Anclam 1867, "Invitation to the public examination of all classes," W. Dietze Anclam.

even more his idea of founding a new solar observatory near Berlin. Wilhelm Förster had originally envisioned a newly founded observatory near Potsdam as a spin-off of his own observatory, because in Berlin "cloudiness from smoke and dust," vibrations due to the poor paving and rising air currents have made the conditions under which the Berlin observatory has to work in the city centre increasingly unfavourable.[25] He was the founder of many institutions and associations among them the famous International Commission for Astronomical Telegram Traffic (1882). At the same time, since 1871 he was the Director of the European Standardisation Commission of the Deutsches Reich and President of the International Committee for Weights and dimensions (Fig. 2.4).

In October 1871, Friedrich had written to the responsible minister that he would personally pursue the *speedy realisation* of such plans. Förster had designed a facility for direct, spectroscopic and photographic monitoring of the sun, which would function simultaneously as a main magnetic and

Fig. 2.4 Wilhelm Julius Förster (1832–1921) in 1913, visionary innovator of solar-telluric physics. (Sculpture by E. Wägener. Photo C.A. Wimmer)

---

[25] Förster (1875).

meteorological station.[26] Behind this was his idea that the central—and for a long time the only—object of the newly developing astrophysics was the sun. On the other hand, he was an early emphasiser of the importance of solar-terrestrial relationships, when he wrote that "the influence of the changing sun on temperatures has so far proved to be small, but should nevertheless provide an important object of research for the future."[27] Förster suggests Spörer (who was completely unfamiliar with weather observations) as the director and himself as a member of a supervisory board. At that time, the Berlin observatory was mainly concerned with orbit calculations of minor planets and comets. Finally, he warned that England already has something similar established for solar and magnetic observations at Kew near London. Förster later wrote about the rejection of his concept, "this was indeed a scope of observational tasks that went far beyond astronomy, in that it simultaneously defined the solar observatory as a meteorological and magnetic-electrical central observatory."[28] No wonder the Academy "shook its head at the whole project, especially since it was a speciality of the older meteorologists that they did not want to know anything about the intervention of the sunspot and flares in the earth weather conditions." He also admitted that the effects he was looking for could be minor and well hidden—so well hidden that such ideas were not taken up again until the next century: in the militarily motivated ionospheric research during the years of the WWII,[29] the establishing of an Institute for Solar-Terrestrial Physics at the Academy of Sciences Berlin and even later in the extensive studies on the influence of cycle-dependent solar radiation on the earth's climate.[30] Nevertheless, almost excessive meteorological measurements were carried out three times a day at 8 different depths during the founding period of the Potsdam Observatory in accordance with the memorandum; they were only stopped (after 17 years) when the newly formed Meteorological Observatory had begun to produce its own data. The printouts of the final 6-year measurement campaign of the AOP contain more than 100,000 three-digit figures on wind, clouds, humidity and above- and below-ground temperatures.[31] The temperature measurements in the deep well were stopped as early as 1888 due to "thermal disturbances," which is why data are only available for a single solar cycle. It is completely unimaginable that, without electronics, Paul Kempf might find the finest effects such

---

[26] Scheiner (1890).

[27] Herrmann (1975) gives the complete memorandum.

[28] Förster (1911).

[29] Seiler (2007).

[30] Solanki & Fligge (1998).

[31] Kempf (1895).

as the influence of sunspot frequency on the ground temperature at a depth of up to 40 m on the Telegraphenberg. He needed own Publications of the Astrophysical Observatory in Potsdam to archive the enormous amount of data.

The commission of the Academy of Sciences, which had to comment on Förster's proposals argued in a different direction. For them, solar research was only the beginning of the new astrophysics. In their letter of 29 April 1872, they supported the solar physics part of the Förster memorandum, but saw it only as part of stellar astrophysics, which had recently begun to develop. A second, independent institute was to be dedicated to "telluric" physics, "in order to meet the needs in the field of astronomy and cosmic physics, … one for astrophysics and the other for meteorology and geomagnetism."

As the two concepts were far apart in magnetism but not in solar research, Förster soon found the solution to end the debate—which he considered hopeless—on 27 May 1873 by naming Spörer for solar research, and in the spirit of the Academy the most outstanding German observer in the field of spectral analysis, Hermann Carl Vogel, astronomer at the private observatory at Bothkamp near Kiel. Everything now happened very quickly. The ministry appointed a founding commission[32] chaired by the Academy President, who soon appointed the building inspector Spieker as a trained architect for public building.[33] As early as June 1873, the concept for an Astrophysical Observatory on the Telegraphenberg near Potsdam was ready, of which solar physics was only a part. After the parliament had approved the financing plan[34] in the winter of the same year, Vogel and Spörer as observers and a little later Vogel's former assistant Lohse as First Assistant were employed—and the road to the Observatory from Potsdam was initially laid out. Spörer moved with his whole family[35] to Potsdam in the same year and continued observing sunspots with his own instruments on a tower in the city until the observatory was completed.[36] Förster described the turbulent founding years around 1874, full of activity at his Berlin observatory and on the Telegraphenberg: Vogel and Lohse are currently in Berlin, where they have also been given rooms for the time being, doing preparations on the large spectroscope by H. Schröder

---

[32] Bois-Reymond (Chairman), Auwers, Foerster, Helmholtz, Neumayer, Schellbach, later Kirchhoff, Siemens and Spieker. Spieker (1879).

[33] Eggers (1995), Pedde (2023).

[34] Bollé (1993): Costs without instruments 860,000 M, equivalent to the costs for the entire Technische Hochschule Charlottenburg.

[35] 3 sons, 5 daughters. According to Wilfried Hornburg (Anklam), all of Spörer's sons (Paul *1855, Richard *1860, Max *1865) attended gymnasium in Anklam just until 1874. The solar observations in Anklam thus ended in June 1874.

[36] Vogel (1895).

Fig. 2.5 Main building on Telegraphenberg in construction 1876. (Courtesy of B. Eggers)

and at the same time, in conjunction with Spörer the care of the facilities of the new observatory (Fig. 2.5).[37]

The Telegraphenberg was just one of several telegraph mountains near Berlin and Potsdam, which carried optical telegraph systems that could transmit messages in a westerly direction from the Sternwarte Berlin to the Koblenz Castle for military purposes. A church was station no. 2, followed by the hill in Wannsee, the Brauhausberg in Potsdam and the Fuchsberg in Glindow, the latter two of which have been called Telegraphenberg since then. The messages were transmitted from station to station using a kind of flag alphabet shorthand, and it was only in the last station that the actual verbose dispatch text 'His Majesty the King ...' was formulated. The system operated from 1834 to 1850 with an average frequency of 2 transmissions per day.[38] The invention and rapid development of electric telegraphy made this elaborate Prussian communication project an episode, and only the name for the location of the Royal Observatories remained.

Gustav Kirchhoff in Heidelberg had declined the directorship in Potsdam, stating that he preferred to develop mathematical physics. He had founded spectral analysis with Bunsen in 1861 and was the first choice of the commission. The two observers were thus placed under the supervision of a technical and administrative directorate[39] until 1881, either because there was

---

[37] Förster (1875).

[38] Arlt (2007).

[39] Auwers (chairman), Förster, Kirchhoff (since 1875 professor of theoretical physics in Berlin).

disagreement or because there were initially doubts about the newly appointed staff. In fact, the imposing Direktorhaus belonging to the observatory was only built later, after Vogel's final appointment as director on 1 April 1882—in a different location to the one intended.[40] Vogel lived there from around 1885 with two housekeepers and three dogs who replaced his missing family. The house had been built according to the specifications for a first-class Prussia counsellor, so it had a representative room with a floor area of 48 m^2, a ceiling height of 4.20 m and a huge double door,[41] in which possibly only a handful official receptions ever took place, e.g., that of the Kaiser Wilhelm for the inauguration of the Great Refractor.

The buildings of the Schinkel school of long-living construction by the architect Spieker are easy to identify. They almost always have yellow brick structures with red stripes and arched windows, occasionally with pillars (Fig. 2.6). He also built a prison, a laboratory on the site of the Universitätssternwarte Berlin, the university library and two institute buildings, the latter almost parallel to the complex on Telegraphenberg. His plan for the Potsdam observatory ground was as simple as it was ingenious. Naturally, the telescope domes for the instruments were to be located at the highest points of the hill[42] and the necessary deep well—an almost invisible masterpiece—including the gas station at the lowest point. Several lockable

Fig. 2.6 Main building of the Astrophysical Observatory from south, 1930. The bricks came from regional factories, foundation walls and stairs from Wefensleben sandstone and the clinker from Siegersdorf/Bunzlau. Middle astrodome 10 m, original status with heliograph in the centre. (Courtesy of AIP)

---

[40] Spieker (1894).

[41] Personal communication B. Eggers.

[42] 64 m above the Havel and 91 m above sea level.

copper tubes for thermometers lead from the 48 m deep well with its surrounding spiral staircase; at a depth of 24 m the shaft leads into an 8 m long test chamber. The well house above ground allowed for pendulum tests. The observatory had its own electricity generator and its own water. The residential buildings for the director, observers and assistants were located halfway up, with separate access and exit roads. The concept also included a fence with dense hedge planting as far away as possible from all the buildings to hamper possible disturbances. The three large domes, which stand in an exact east-west direction, were to be located at some distance from each other, were to be accessible from one another on several levels, on the lower level through walkways with open round arches. They form the south wing of the main building, for the construction of which a massive earth wall of a forgotten defence line from 1813 had to be removed first. The north wing is centred along the meridian and contains three huge offices on the main floor, connected to a modern air heating system. The north entrance is crowned by an imposing water tower with an internal iron spiral staircase, visible from afar, which is accompanied by a similarly designed, slightly minor tower of the Direktorhaus.

Triumph with the Demon Star

The water supply and the residential buildings for the observers, the assistant, the institute's servants and the machinists were completed at the end of 1878. First Spörer then Vogel, Lohse, the servant Doll and the machinist Meier had moved in. The construction of the main building was behind schedule, but parts of the building complex had already been in use since October.[43] The entire building was only completed and handed over in the spring of 1879 (Fig. 2.7). The three scientists appointed Vogel as their speaker and Spörer as the person responsible for the instruments and books, certainly a preliminary decision on the future structure (Fig. 2.8). Spörer "endured all the mutations of life with an indestructible calmness, and always soon returned to business as usual, i.e., to his studies of the sun,"[44] is how Lohse describes the atmosphere on Telegraphenberg. From the summer of 1877, Vogel had Gustav

---

[43] Spieker (1879).
[44] Lohse (1895).

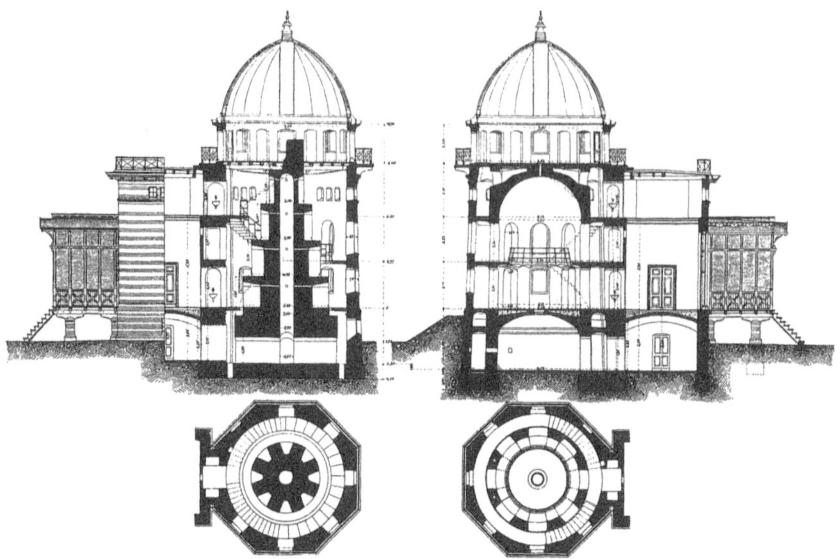

Fig. 2.7 Western astrodome (left, own telescope fundament) and eastern astrodome (right, no telescope fundament), both 7-m-domes were initially used for solar observations. In the lowest round room under the east dome (due to the particular rigid masonry), A. Michelson operated his famous experiment which ruled out the existence of a stationary ether (Shankland, 1982, originally the experiment has been considered as failed). Michelson was enrolled at Berlin University at the time, the interferometer was built by Schmidt & Haensch in Berlin (Scheiner, 1890; Bleyer et al., 1979; Auth, 1982). (Source: Scheiner, 1890)

Fig. 2.8 First residents of the Telegraphenberg, left to right: G. Spörer (1822–1895); H.C. Vogel (1841–1907); O. Lohse (1845–1915). (Courtesy of AIP)

Müller[45] as his personal assistant, with whom he had previously worked at the Sternwarte Berlin, and after a further year, Paul Kempf, also from Berlin.

For the work of Professor Spörer an assistant has not yet been employed in the past year, wrote the directorate in the 1877 annual report to the Minister of Education. In the half-completed institute, he continued to do what he had already done in Anklam, still using his own instruments in the East dome. Already in November 1870 he had received a spectral apparatus in order to be able to see the solar prominences in the light of their yellow line, but only made a few observations in winter, although the sun was at its spot maximum. For the first time his own prominence observations happened during fall 1871: Flaming prominences followed September 6 southeast of the location of the minor spot and reached as far as the area of the following group.[46] He observed the sun on 229 days in this year, whereby it remained spotless on 103 days. The heliographic coordinates were calculated and averaged for the spots registered in projection. The spot zone had always slowly moved equatorward over the course of the 11-year cycle. The year 1877 was unfavourable for the observation of prominences. A single prominence, observed in June was reported in detail,[47] later hundreds of prominences were registered annually in the eastern dome, but hardly ever analysed. Still from Anklam, in 1871 Spörer had published his algorithms for determining the heliographic coordinates, in order to enable observers to make a comparison with his observational results. For demonstration, the positions of some prominence observations in Leipzig were calculated. He was actually only interested in the feature of prominences occurring also at much higher heliographic latitudes than the sunspots; with sunspots, the law of rotation could only be determined for lower heliographic latitudes.

Since 1881, the southern stem of the middle dome had housed a huge heliograph, the largest of its time with an aperture of 16 cm and a focal length of 4 m, in order to be able to image the solar disc with exposure times of thousandths (!) of a second (Fig. 2.9). "The idea that a fixed installation of the heliograph parallel to the Earth's axis would offer great advantages, tempted Prof. Vogel and myself to work out a project for a heliograph in 1874, which later formed the basis for the realisation of the Potsdam instrument."[48] The Potsdam heliograph is also similar in details to the apparatus of de la Rue and

---

[45] Director of the observatory 1917–1921, married to Marina Luise Spörer (1853–1885); their daughter Gertrud became the wife of G. Eberhard, his niece Käthe Müller married H. Ludendorff.

[46] Spörer (1873).

[47] Astron. Nachr. 90, 63 (1877).

[48] Lohse (1889). The year 1874 in Lohse's description of 1889 may not be correct.

Fig. 2.9 Left to right: the 30-cm-refractor by Schröder for spectroscopy, Vogel's main instrument; the heliograph installed in the southern porch of the main building. The heliostat, which is fixed to the outside ground, reflects the sunlight from below into the 4-m-tube, a rigid camera is in a separate darkroom. (Courtesy of Förderverein Großer Refraktor Potsdam)

J.N. Lockyer which Vogel had seen in England in 1875 and described in detail in his travel report. The sunlight was reflected by a heliostat into the optical system from below and finally reached the very heavy camera at the top (Fig. 2.10). Vogel was convinced—probably in contrast to Spörer—that the traditional projection method was only sufficient for simpler questions of spot statistics, but that more precise findings, e.g., about the proper motions of sunspots, could only be achieved photographically. For stellar spectroscopy only later did Vogel favour the photographic plate.[49]

On his trip to England, Scotland and Ireland[50] in the summer of 1875, Vogel had seen a "photoheliograph" in Greenwich, the existence of which may have confirmed or even generated his thinking. Vogel had been guided by the photographic assistant E.W. Maunder through the observatory and had certainly noticed that he had only recently been employed to monitor the sun daily with the heliograph. The solar images had a diameter of 10 cm, the

---

[49] Lohse (1907).

[50] Gußmann & Dick (2000). On this trip Vogel saw the giant telescope (refractor of 63 cm aperture and 9 m focal length) of the private observatory of R.S. Newall near Newcastle and the "Leviathan of Parsons town" (mirror with a diameter of 183 cm and a focal length of 16 m) in Birr near Dublin.

Fig. 2.10 Heliostat as part of the flagship of the observatory from Steinheil (optics), Repsold (mechanics). With a similar instrument made in 1890, Schwarzschild attempted to measure the gravitational redshift of the nitrogen line 3883 Å on the roof of the Beamtenwohnhaus in 1913. (Source: Lohse, 1889)

plates were developed wet, but their quality left much to be desired: "The detail, even in the spot groups [of 1869 and 1870] was only slight." In Potsdam, Lohse began taking regular photographs of the sun already in July 1882 (Fig. 2.11). He also experimented at this time with perforated photo plates and special emulsions in order to be able to image the chromosphere and corona without the central sunlight, which outshines everything else. At times, the observatory on the Telegraphenberg must have depicted a research centre for photography with several, according to Vogel's command, extremely clean darkrooms. However, Vogel later realised that "the optical performance of the new Potsdam instrument did not match the expectations, and the observer had decided to use only a minor central part of the objective." The south exit of the main building, which emerged from the dismantling of the old heliograph and serves as a convenient connection to the southern institute buildings, was only installed in 1960.

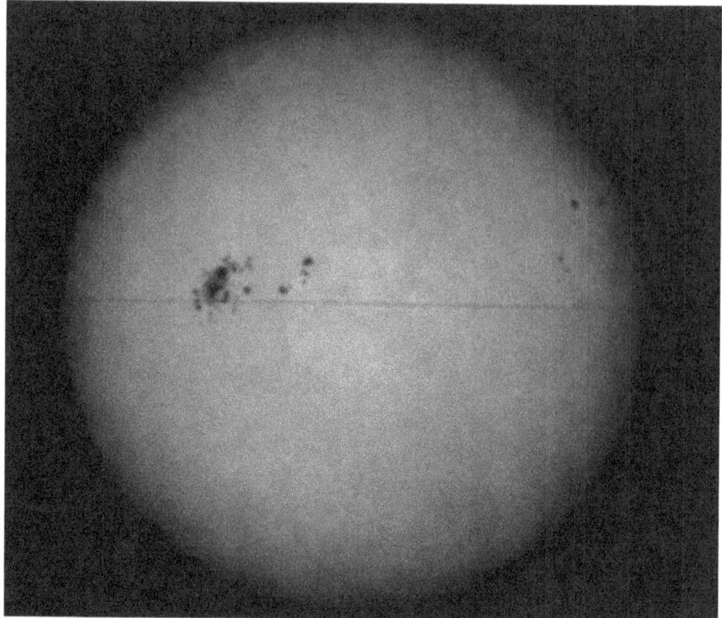

Fig. 2.11 Heliograph plate taken on 13 February 1892 as an example of the daily registrations by O. Lohse. Note the limb darkening independent of the latitude. "The heliograph has so far only been used for small recordings" (Scheiner, 1897, plate III)

In 1899, a 16 m long spectroheliograph was established on the roof of the Beamtenwohnhaus, which was built at the same time as the Great Refractor building and contained a spectrograph in the southern tower (Fig. 2.12). The actual heliostat in front of the southern outer wall of the northern clock room reflected the sunlight from the north into the optical system. This meant that the sun could be observed in monochromatic light with a grating spectrograph, first visually and later photographically. Vogel had originally intended to have the facility built at ground level west of the main building and had already ordered the heliostat mirror in 1890, but when the construction[51] of the Great Refractor and the associated offices actually came about, he simply placed it on the roof.[52]

Vogel had also visited the mechanic H. Grubb in Dublin 1875 to discuss the instrument he had ordered for Potsdam. He found the opportunity here to gain "a wealth of experience and learn a lot from the intelligent young man." The vibration-free West dome will later house this Grubb refractor

---

[51] Architects: Paul Spieker (1826–1896) and Eduard Saal (1860–1918). See Saal (1901).

[52] Due to the roof construction, the heliostat track was distorted by air turbulence and heat development despite numerous corrections.

Fig. 2.12 Beamtenwohnhaus (= library building from 1959), built in 1898/99 for the observer and the machinist of the Great Refractor. Roof structure: spectrograph room with extra insulated walls (left), clock room (right, lost). The heliostat was located on the southern outer wall of the clock room (Saal, 1901)

with an aperture of 20 cm[53] and a focal length of 340 cm for observing the sun and stars, which Spörer used from 1879 onwards and which led to sunspot drawings of the highest quality every clear afternoon (Fig. 2.13). How quickly the spots disappear after their appearance had not interested him and many of his successors, although Vogel had seen that in London "the area that the sunspots occupy on the solar surface was also deduced from the … photographs." The rotational behaviour of the bright chromosphere flares has also rarely been considered. The bright faculae visible on the Potsdam plates have only been measured in exceptional cases. Unexpectedly, the law of rotation of the flares calculated by J. Wilsing proved to be much flatter than that of the sunspots,[54] which, however, did not remained unquestioned.[55] Even the poleward meridional flow and the correlation of the latitudinal and longitudinal proper motions of the larger groups of spots will have certainly been included in Spörer's data but were only discovered almost 100 years later.[56]

From 1889 to 1945, a very special double refractor with a 32 cm photographic lens and a visual guide tube with an aperture of 24 cm stood in the

---

[53] Since 1908 new lens with 30 cm aperture by Steinheil.

[54] Wilsing (1888).

[55] Belopolsky (1893), Kempf (1916).

[56] Ward (1965).

Fig. 2.13 Spectroheliograph on the Grubb refractor, part of the instrument on the roof of the Beamtenwohnhaus since 1908 (Hassenstein, 1941; Kempf, 1905)

Bekanntmachung.

Die Besichtigung des Königl. Astrophysikalischen Observatoriums ist dem Publikum ieden Freitag von 3 bis 6 Uhr
nachmittags unter folgenden Bedingungen gestattet:

Fig. 2.14 "Visits to the Royal Astrophysical Observatory are open to the public every Friday from 3 to 6 pm under the following conditions…" Regular public relations were expected of state scientific institutions in Prussia (Förster, 1911)

Photo dome erected in the west of the main building on the Telegraphenberg (Figs. 2.14 and 2.15). It served the first international astronomy project of a complete photographic sky map with 40 million stars up to the 14th magnitude and a general catalogue up to the 11th magnitude with about 2 million stars (Carte du Ciel, CdC), whereby the strip between 32° and 39° northern latitude was assigned to the Potsdam observatory. No other German institute

Fig. 2.15 Left to right: the Photo dome with darkroom, completed in 1889 for the Photographic Sky Map initiated by the Paris Observatory. Vacant since 1945; astrograph by Repsold (mechanics) and Steinheil (optics) for the international project Carte du Ciel. The special mounting of the double refractor allowed convenient handling for observations close to the zenith (Vogel, 1907)

had participated in this originally French program, which was developed at an astrophotography congress[57] in Paris in April 1887: American institutions also did not show interest. Because of Lohse's good experience with solar photography, Vogel had gladly accepted the invitation to the congress and even waived the approval of ministerial travel funds. The AOP, as the best-equipped German institute, turned down the sky map because of the excessive exposure times but wanted to work intensively on the Astrographic Catalogue. Immediately after returning from Paris, Vogel applied for the necessary 52,000 marks to build the facility and also for the funds for an additional observer.

Just 10 years after the photo dome went into operation, Scheiner—who was still best friends with Vogel—published apparent orthogonal coordinates of stars up to 11th magnitude for the Potsdam zone. The 2002 images, obtained from 1889 to 1923 by Eberhard and Ludendorff under Scheiner's direction—later by Münch and Birck—were stored in the building of the Great Refractor, but only 977 of the 16 × 16 cm plates survived WWII. The war's consequences as well as improper handling,[58] including all the plates of the second series from 1913–1924, which had been initiated by Karl

---

[57] Participants from Berlin/Potsdam: Auwers, Lohse, Vogel. Förster had cancelled (Lamy, 2008).

[58] Most of the photographic plates from the series 1893–1900 have been lost. The instrument was removed in 1945.

Schwarzschild for the derivation of stellar proper motions in identical star fields are all lost to time.[59] The astrometric work, however, was completely stopped at the end of 1924 because the interest in catalogue work coordinated by Paris was extinguished. The Observer Otto Birck retired in mid-1924 at the age of only 45 by the newly appointed director Ludendorff. Vogel had already noted that the view [was] prevailing that this double refractor was a completely perfect instrument in terms of its mechanical design and optical performance, which he had increasingly preferred to use for spectroscopic investigations rather than for astrometric tasks.[60] The production of the sky map was only remotely related to the actual field of the institute´s activity and should therefore "stand in the background of the other work."[61]

In 1908, G.E. Hale from the Mount Wilson Observatory in California concluded by comparing the spectral lines of sunspots and their surroundings with his spectroheliograph that sunspots are a magnetic phenomenon.[62] Hale had started to answer Carrington's old question "What is a sun?" suggesting that it is a magnetic star, too. This pioneering work, however, did not trigger any demonstrable reactions on Potsdam-Telegraphenberg despite of the hints for chromosphere activity of a few stars—perhaps because it was the period between the finished Vogel- and the coming Schwarzschild era after the old director and successful science manager[63] had died the previous year. The heliograph in the main building had continued to operate routinely in white light without visible results, the new roof instrument was equipped in the same year with the instrument by Kempf[64] which had been probed in the west dome. It is questionable whether the actual solar researcher in Potsdam did realize the new challenge to think in terms of magnetostatics and electrodynamics. Polarimetry—the magnetically-splitted (Zeeman) lines are circularly polarised—was not mentioned in the AOP reports for a long time; this deficit was only overcome 33 years later by Harald v. Klüber in the Einstein tower telescope in 1941.

Karl Schwarzschild had used the roof heliograph in 1913 for his search for Einstein's gravitational shift of light wavelengths; earlier publications with data from this instrument have not appeared, the study by Kempf on the rotation of the calcium network—with material from a 6-month observation campaign in 1906 with the Grubb refractor in the west dome—was only

---

[59] Dick (1988).
[60] Scholz (2000).
[61] Bigg (2008).
[62] Hale (1908).
[63] Real cost of the Great Refractor: 706,250 marks, including 270,000 marks for instruments.
[64] Kempf (1905).

published in 1916.[65] Kempf's results do not provide any proof that calcium, flares and spots provide different rotation laws. Unfortunately, "it was not possible to work on the rich material obtained so far, as there was a lack of a suitable assistance," complained Director Vogel typically and regularly in his annual reports from 1904 and later.

The grand instrument with an aperture of 30 cm and a focal length of 540 cm for Vogel's spectroscopic work, called the "great refractor" (until 1899), was part of the middle dome and also stood on its own footing separate from the building shell. The construction led to the creation of an impressive cupola room on the working floor of the building. Even before his time in Potsdam in the private observatory Bothkamp (1870–1874, see Fig. 2.16), Vogel had shown that the different rotation of the edges of the sun could be

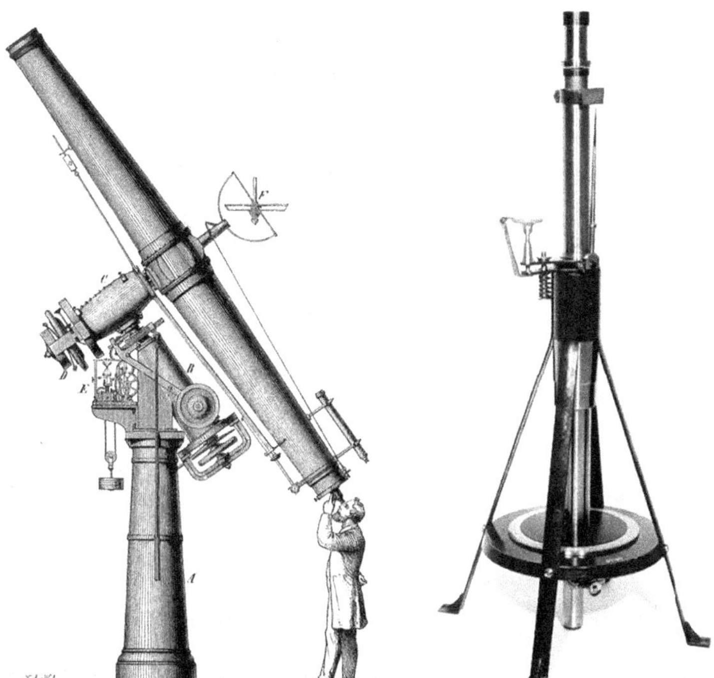

Fig. 2.16 Vogel's instruments both built 1871 by Schröder in Bothkamp where he successfully probed the Doppler line shift due to the solar rotation. Left to right: the then largest telescope in Europe; the vertical spectroscope. They both mark the beginning of modern astrophysics. (Courtesy of Deutsches Museum München/Rainer)

---

[65] Kempf (1916). The daily monitoring of the solar chromosphere in the light of the calcium K-line by the Mt. Wilson Observatory started 1915 and ended 1985.

detected using visual spectroscopy, hence the Doppler effect also works with light. This was no minor discovery. He wrote, "that the observations always give a greater velocity than that deduced from the known rotation time of the sun … but it emerges from all observations, that a displacement of the lines, caused by the rotation of the sun, can be regarded as proven."[66] He was, however, sceptical about applying this concept to stars, arguing that there were simply no stars whose lines were all measurably broadened. However, if the frequently encountered broadening of the hydrogen lines were interpreted as a Doppler effect, Vega, for example, would rotate at more than a hundred times the surface speed of the sun. The use of photography, he assumed at this time, is not expected to be significant either, because visual observations would be far superior to these. However, the spectral lines of stars are so weak in visual observations that valuable results could only be expected for the Doppler shifts using powerful instruments.

Vogel had then attended the first astrophotography congress in Paris 1887 and drastically changed his mind there about the application of photography to spectroscopic investigations.[67] In the following year he was already able to increase the absolute accuracy of his results with photographic methods to around 7 km/s and demonstrated the radial movements of Sirius, Procyon, Rigel and Arcturus. The limit of a Doppler shift of 7 km/s results from the restricted ability to measure 0.01 mm on a photo plate by means of a microscope (Fig. 2.17). In his publication of 1892 about a list of proper motions in the line of sight of 51 stars one finds today only 6% wrong signs compared with the modern results.[68] His method worked, but of course with limited accuracy only.

The next year provided an unexpected sensation. The hydrogen lines of the bright star Spica compared with lines generated in Geissler tubes inside the telescope were differently shifted on different days. Spica was now observed almost daily. The radial velocities showed a sinusoidal curve with an only 4-day period and maximum values of 20 km/s. Vogel assumed to see two stars with solar mass and only one (!) solar radius apart and with great vision he predicted "that there will be a large number of close binary star systems; the world of orbital periods will range from many centuries down to a few days and of tiny distance until the atmospheres touch each other."[69] Spica was the first known binary whose individual stars could not be resolved even with the

---

[66] Vogel (1872/1873).

[67] Lohse (1907).

[68] Vogel (1892).

[69] Vogel (1890a), Rigel is also mentioned in this publication.

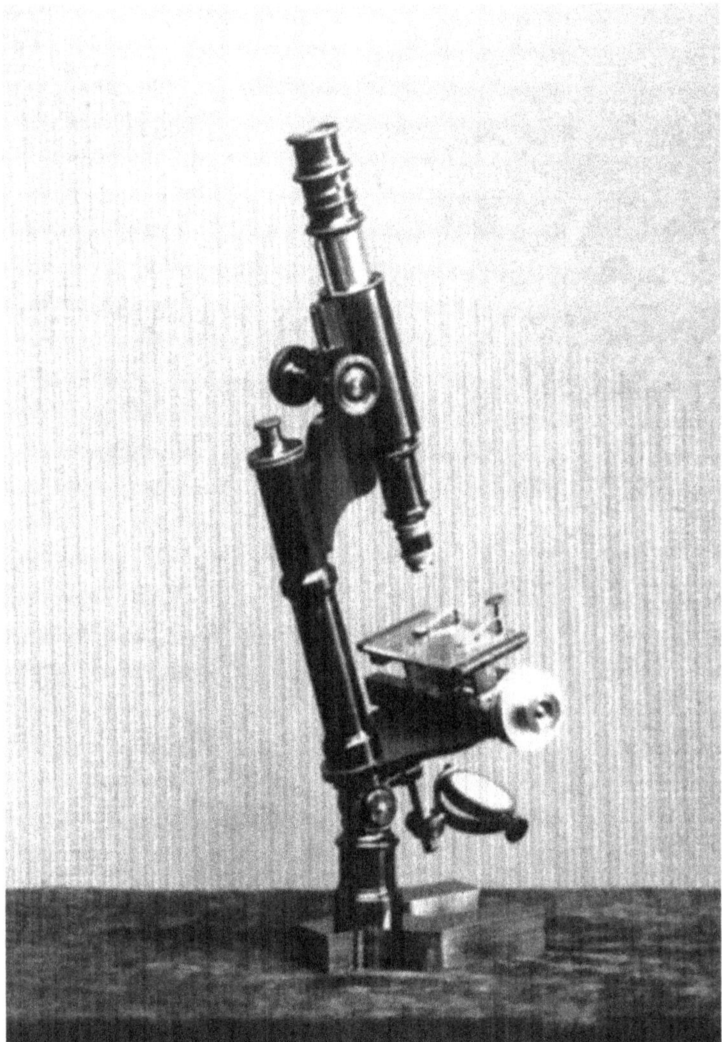

Fig. 2.17 Vogel's microscope to measure Doppler line shifts on photographic plates (Scholz, 2000)

largest telescopes of the time. The following winter, spectral images of Algol known as the "demon star" were obtained, which showed that Algol moves away from the sun before a minimum, and approaches it after the minimum, "as must be the case when a dark body moves in front of Algol."[70] In the past, visual observations had always remained inconclusive so that there was no

---

[70] Vogel (1890b).

scientific explanation for the rapid change in brightness visible to the naked eye. The spectroscopy had celebrated one of its greatest triumphs[71]—to attend the photographic congress in Paris was the decision of his life. Vogel now wanted more. In a memorandum addressed to the ministry in 1890 entitled "On large telescopes and their significance for science" it was stated that "for the further training of astrophysics—the main task of the Astrophysical Observatory in Potsdam—it will be necessary to provide the institute with a larger instrument, which alone would enable us to maintain the highly respected position it had acquired over the 16 years of its existence, especially abroad."[72]

Vogel was the appointed director of the observatory from the beginning of 1882, Spörer became the First Observer and the external Board of Directors ended its activities. Now the "Sonnen-Warthe" quickly developed into the "Astrophysikalisches Observatorium zu Potsdam" (AOP) and the annual reports were primarily devoted to the results of spectral analyses of various celestial bodies, with the sun only being reported in the last sentences. In December of the first year under Vogel, Spörer wrote an alarming letter to v. Helmholtz in his role as President of the Physikalische Gesellschaft zu Berlin that he had wanted to come to the regular meeting of the society in Berlin *last Tuesday* to present his observation of the Venus transient, but Vogel had forbidden him to do so and declared his result to be physically misleading. Every lecture and every publication would now have to be authorised. Spörer described his problem to Helmholtz in great detail because he feared *that the ban would also extend to my solar investigations and that I would be prevented from publishing anything that did not completely agree with the director's views.*[73] The collegial relationship between the two astronomers survived Vogel's promotion only by a few months; the new director (and newly appointed professor) found the right to issue instructions to the employees too tempting. Nevertheless, a second assistant position was budgeted from 1886, which was given to Paul Kempf so that there were now four regular posts in addition to the director.[74] Kempf's determination of Jupiter's mass as 1:1047.7 of solar mass is still valid today.

---

[71] Müller (1907).

[72] Vogel (1907).

[73] Spörer's letter to Hermann von Helmholtz dated 21 December 1882, GStAPK.

[74] Status 1890: Principal Observer: Spörer; Observers: Lohse, Müller; Assistant: Kempf, Wilsing, Scheiner. Kempf became Observor in 1894, Scheiner became Principal Observer in 1898, Wilsing received a full assistant position 1895.

In the end, as is often the case, things became uncomfortable. In a detailed status report of all scientific institutions on the Telegraphenberg[75] from 1890 for the public, it was stated, signed J.S., that "the systematic observations of sunspots, which have continued for more than 100 years … had not opened up significantly new perspectives in recent decades" and that "interest in sunspot statistics had reduced more and more," as the spectral analysis was the observatory's main field. Was this just a professional misunderstanding or was it coolly aimed at the future retirement of the 68-year-old Spörer? It is unimaginable that Vogel would not have seen this affront of his First Observer before it went to press—he may have ordered it personally or even pre-formulated parts of it. Vogel had often described himself as the First Observer of his institute, although after his appointment as director he only occasionally stood at the telescopes himself. The rest of J.S.'s description of the institute is written in a well-informed, detailed and directorial manner—certainly not the level of a newcomer to the observatory. It was only at the beginning of 1887 that Julius Scheiner, who had come to Potsdam from Bonn through intervention of Vogel's sister, had quickly become Vogel's favourite student and often had to represent him—mostly for health reasons—at conferences and other official events (Fig. 2.18). At the ceremonial inauguration of the Great Refractor on 26 August 1899, after the director's speech, Scheiner was allowed to explain the imposing instrument and dome to Kaiser Wilhelm, who was present. The friendship between Vogel and Scheiner ended soon afterwards due to both of them despairing over the poor optical quality of the instrument and diverging opinions on the cause of the disaster,[76] possibly due to the suspicion that they had jointly overtaxed the technology. The announced revolution in stellar physics based on powerful lenses would not happen, at least not in Potsdam, and to get new lenses was not possible. In 1903, the director explained that "the Principal Observer at the Astrophysical Observatory in Potsdam, Prof. Scheiner has published a communication which he failed to submit to me before sending it to the editors of Astronomische Nachrichten."[77] Scheiner had, perhaps wrongly, criticised his colleagues Hartmann and Eberhard of not having been aware of published observations of arc spectra, but the background was that the ex-friend had omitted "to submit to me." The later director Gustav Müller reported that Vogel liked to avoid close contact with the scientific staff at the observatory. There were no joint meetings and consultations on important matters concerning the

---

[75] Scheiner (1890).
[76] Wilsing (1914).
[77] Astron. Nachr. 162, 159 (1903).

Fig. 2.18 Julius Scheiner (1858–1913). (Courtesy of Förderverein Großer Refraktor Potsdam)

institute, and it was only on very rare occasions that the entire staff gathered around him. He preferred to negotiate with everyone alone, and for a long time he only communicated with some of them in writing.[78]

The history of the Great Refractor and its building began in 1890 with Vogel's memorandum "On large telescopes …" and a trip to England. He had regarded the construction of the Great Refractor mainly for the continuation of spectral-analytical research and here especially "for the performance of velocity determinations of the stars in the radius of vision" as his life work; he survived its realisation by only a few unhappy years. Everything here, the rotating giant dome, the precise telescope tracking, the functions of the guide

---

[78] Müller (1907).

tube, the sophisticated mechanics of the observer's seat and even the modern gas-driven dynamo apparatus stood and fell with the quality of the large objective. It was the greatest catastrophe imaginable, Vogel hardly ever entered the dome of the Great Refractor (Fig. 2.19), working—if at all—with his old

Fig. 2.19 The Great Refractor from 1899, dome 21 m, astrograph 80 cm, visual guide tube 50 cm, focal lengths about 12 m, Repsold (mechanics), Siemens & Halske (electrics, gas-powered dynamo). The floor is rigid, the observation chair is attached to the dome. After Vogel the optical 50-cm-lense proved to be excellent during the examinations. The astrograph was the problem: Steinheil tried to save the 300 kg lens in Potsdam in May 1900 and in Munich in January 1904 without success. Photo ca. 1964. (Courtesy of Förderverein Großer Refraktor Potsdam)

refractor in the middle dome of the main building whose lack of light intensity he had complained about in his memorandum. A good unification of the light rays can only be achieved with all larger lenses by means of retouching due to the inhomogeneities of the glass,[79] he took heart himself. The lenses showed spherical aberration, i.e., their circular zones produced separate focal points.[80] Wilsing and Scheiner had roughly criticised even after the optical improvements that the objective "is now considerably inferior to what it was before, and that in its present condition it is not suitable for finer observations." Vogel replied that there will be nevertheless an almost inexhaustible field of work in which the 80-cm-refractor in its present condition will prove its worth in the hands of skilful and clever observers.[81] The comment is aimed at future investigations of visual binaries. In this publication, Vogel also describes in detail the development of Hartmann's optical procedures, which later became famous for testing large lenses.

The Aged Spörer Finds the Maunder Minimum

At the end of the 1880s, the old Spörer discovered the today well-known Maunder minimum of solar activity, the significance of which was not only unknown to his closest colleagues. Spörer had, in addition to his own extensive archive of sunspot registrations, obtained all the documented observations from Rudolf Wolf in Zurich in order to trace his passion—each solar cycle begins with spots that appear at high heliographic latitudes and ends with new spots close to the equator—back to the beginning of the records in 1618. To his surprise, this only worked until the minimum in 1713, because there were no spots before at all in the northern solar hemisphere for decades, and the few southern examples only ever appeared close to the equator. More normal conditions were only found again in the records of Christoph Scheiner for which he for 1621 found latitudes of 27° and for 1626/1627 latitudes of 10°—according to his expectation—but all the spots were located in the southern solar hemisphere. In August 1887, Spörer announced at the annual meeting of the Astronomische Gesellschaft his finding that since the middle of the seventeenth century, over a very long period of time, conditions on the sun have been substantially different from those in more recent times.[82]

---

[79] Vogel (1904).

[80] Up to 6 mm difference for the 80-cm-lense.

[81] Vogel (1907), Hartmann (1908).

[82] Spörer (1887) and Nova Acta der Ksl. Akademie der Naturforscher Leopoldina, vol. LIII. No. 2, Halle (1889).

Edward Maunder from the Royal Greenwich Observatory had presented in detail the fundamental theorems of Spörer's publication of 1887—that from the middle of the seventeenth century until 1713 (1) there were no sunspots far from the equator, (2) that only one hemisphere was spotted and (3) spots were very rare—to the readers of the Monthly Notices in 1890. Three years later, in an illustrated science magazine, Maunder himself placed a short and entertaining story of historical spot records from a large number of solar observers and—since he already knew the facts from Potsdam—very quickly came to the conclusion that there had been a "prolonged sunspot minimum" before 1715.[83] Spörers name appeared only once in the references but not as the discoverer of the Grand Minimum. Almost a 100 years later, when it had become a focus of the dynamo theory of stellar magnetic fields, it was suddenly called the Maunder Minimum.[84]

Spörer devoted the rest of his years to the visible differences in activity on the two solar hemispheres with growing astonishment. "In addition to the photographs taken by Dr. Lohse, my own photographs are also used to calculate the heliographic positions. " The measurement of the plates was carried out by Wilsing. He only returned to his most puzzling and now most famous result at the very end—possibly because of the insults of J.S. & Co. "Various reports agree that from 1645 to 1670 there were only a few spots. The number of spots then increased, but there was definitely no significant maximum until 1716." He concluded that in the period of 70 years in the northern hemisphere there was absolutely no periodicity of the spots,[85] hence the Schwabe cycle had been stopped. In their obituaries for Spörer, Lohse and Vogel completely ignored his main discovery; they probably did not take the matter too seriously. Also, the obituaries of Maunder do not contain references to a minimum discovered by him. And so, it came about that Carrington's spot zone migration was named after Spörer and Spörer's long minimum were named after Maunder—a wrong but well-balanced German-English discovery history.

Maunder had complained in his report that no magnetic data exist from the reign of the Sonnenkönig Louis XIV from 1643 to 1715. Indeed, the answer to the underlying question of whether the cyclical magnetic influence on the earth had also disappeared with the sunspots would have allowed speculation about the possible magnetic nature of the sunspots. The next question, whether the magnetic clock in the sun had continued to run with

---

[83] Maunder (1890, 1894).

[84] Eddy (1976), he named Spörer's discovery "Maunder-Minimum," allegedly because of the alliteration.

[85] Spörer (1894).

reduced amplitude during the long minimum or had come to a temporary standstill, could have been asked even then, but interest in this discovery quickly died out because it was more convenient to assume that there were too few active observers at the time to register all the existing sunspots. Prof. Spörer had completed the long series of his work on sunspots and had "retired from my staff on 1 October 1894 at the age of 72," the Director Vogel informed in a preface of the Potsdam publications of 1894. He had entrusted the Observer Kempf with the continuation of the sunspot statistics but such observations no longer took place. Among the numerous publications by Kempf there are none on sunspots, apart from a report from 1910 on "rotating sunspots," which, however, is based on Lohse's photographs up to 1893. No favouring of a sense of rotation on either hemisphere was found. Solar vortices and magnetic fields were Hale's favourite topics, too.

Julius Scheiner received a permanent position and the appointment of associate professor at Berlin University on the same day.[86] In the annual report for 1903, Vogel reports only 38 regular entries for spot observations. The original solar observatory had finally become a famous institute for stellar physics. Whether Spörer was still allowed to use his study for some time, we do not know. But the well-deserved retirement that had been announced did not appear: after just 9 months, on a Sunday trip to Giessen, the "true Berliner," who had always been healthy, died of cardiac arrest. His then 35-year-old son Richard had just taken over the management of Georgshütte Burgsolms, part of the Buderus company. The father had probably wanted to witness the ceremonial inauguration.

---

[86] Salaries 1897: Vogel 10,500 RM, Observers: Lohse 6500 RM, Müller 6000 RM, Kempf 5100 RM; assistants: Wilsing 4260 RM, Scheiner 3260 RM. Informations by K.-D. Herbst.

Part II

A Long Run to the Magnetic Sun

3

Einstein's Tower and Miethe's Dome

From Roof to Roof

The ingenious Potsdam pioneer of colour photography and successful writer Adolf Miethe was appointed to the Chair of Photochemistry and Spectral Analysis at the Hochschule Charlottenburg on 1 October 1899 (Fig. 3.1). To develop 3-colour astrophotography, he had built a photographic observatory

Fig. 3.1 Adolf Miethe (1862–1927), (Miethe, 2012)

G. Rüdiger, *The Astrophysical Observatory Potsdam - Triumph and Tragedies*, Astronomers' Universe, https://doi.org/10.1007/978-3-032-06294-9_3

on the roof of the institute building although the installation caused many difficulties. Ultimately, he was able "to return to the old love of my life: astronomy. A favourable opportunity provided me with this new possibility." This opportunity was a telescope with an aperture of 30 cm made by Gustav Heyde in Dresden which was for sale under fair conditions. His minister gave permission in order to promote photographic astronomy at his chair. After brief negotiations, the transfer of the instrument and its 6 m rotating astro-dome to Charlottenburg was done. "My joy was indescribable."

Initially Miethe made his own astronomical experiences on the Telegraphenberg observatory in connection with the Astronomy Congress in Paris 1887 where "the plan to create a large photographic map of the sky was adopted also had an impact on me. Lohse had intended to present the open clusters he had photographed with the Potsdam refractor. However, some illness delayed his work and he asked me to come to the Observatory and, as an assistant, to carry out a series of measurements and drawings based on his photograms. Nevertheless, I did this simple work with great enthusiasm and to his satisfaction, while at the same time using the Observatory's resources to complete my doctoral thesis. On his advice, I chose to work on the basic photochemical law, which … essentially stated that the photochemical power was proportional to a product of exposure time and light intensity."[1]

The observatory, however, had not impressed Miethe too much. He noticed that Spörer could not name any lunar crater. In the winter 1887 he went to Göttingen, "where I mainly intended to perfect myself further in theoretical physics, especially in optics, and in astronomy."

On the anniversary of his personal observatory in Charlottenburg, he produced his first colour photographs of the moon using the new instrument. Miethe met the optician Bernhard Schmidt[2] from Mittweida shortly after observing Halley's comet with his refractor: I was sitting in my study when a tiny one-armed young man introduced himself and asked to demonstrate a reflecting telescope he had made. He had the telescope with him in a minor box, from which, looking around next to the 30 cm diameter mirror, a very simple construction of a few wires and wooden rafters developed, which he set up in our dome. …One look through the slatted frame convinced me that I was standing in front of a miracle of optical art that considerably dominated my large telescope in terms of real performance.[3]

After he had mounted Schmidt's mirror to his telescope he noted that with the new mirror, "my large instrument only became the guide telescope." Now

---

[1] Miethe (2012).
[2] Schorr (1936).
[3] Miethe (2012).

Fig. 3.2 First colour photograph of the moon, fully mechanical two-colour print, 30-cm-mirror telescope

colourful images of the moon were possible (Fig. 3.2). In front of the focus were the absorbing cuvettes, one of which only let through light with a wavelength of 360–330 nm, the other only light with a wavelength of 700–600 nm. Miethe enthusiastically reported in April 1911 to his ministerial director: We have actually discovered the possibility of geological observations on the lunar surface.

Adolf Miethe was not yet satisfied with this. In order to deepen the results, a much larger reflector telescope was constructed, the optical parts of which were supplied by Schmidt and the mechanical parts were manufactured by Goerz (Fig. 3.3). It is still one of "the most perfect instruments of its kind."[4] It was the first astronomical instrument that the Goerz AG (later Zeiss Ikon) produced in Berlin. Adolf Miethe was a member of its board. Additionally, a 40-cm-reflector[5] was built in 1913 especially for the joint solar eclipse expedition of the Technische Hochschule Charlottenburg and the Goerz AG next year to northern Norway. Miethe had indeed achieved a first co-operation between his institute and a commercial company. As a result, the expedition had a brilliant set of equipment packed in 58 boxes. The main aim was to take corona images using the 30-cm-mirror from the guide tube of the

---

[4] Miethe (2012).
[5] Now: Deutsches Museum München.

Fig. 3.3 Goerz-Schmidt-double reflector (30-cm-mirror plus 50-cm-mirror), until 1945 the key instrument in the Miethe dome on the Telegraphenberg (today A23). (Courtesy of G. Kühn)

Goerz-Schmidt-reflector, whose focal length was enlarged from 7 m to 20 m with horizontal storage, which enabled a solar image with a diameter of 20 cm.[6] The fate of this and the other expeditions—Miethe was the only one who refused to go to Russia—was not disturbed by the weather this time, but by the outbreak of the WWI. All the younger German members of the expedition were ordered back leaving Miethe by himself. He experienced the solar

---

[6] Miethe et al. (1916).

eclipse together with his massive luggage in the most beautiful weather almost alone. Apart from the three boxes he had brought with him, his instruments only arrived back in Berlin many years later. The activities of all the other expeditions to Russia were completely prevented by the war. The Frankfurter Zeitung in June 1916: Contrary to the Russian hospitality first offered, the members of the expeditions who were required to serve in the German army were taken prisoner and could only be exchanged after a long time. In preparation for the 1887 eclipse expedition of the AOP to southern Russia, Miethe had learnt the Russian language "with absurd effort at the request of the Potsdam astronomers in half a year. … When one day I was told with a cold smile that they had decided otherwise, I thought I had the right to be terribly angry."[7]

Miethe's reflector consisted of two high-quality Schmidt mirrors of 50 and 30 cm aperture with 300 and 180 cm focal length, the mechanics made by Heyde. Special mirrors allowed the focal lengths to be extended to 11 and 21 m, respectively; the ocular of the minor mirror and the cassette of the other were mounted together. However, the instrument was used only seldom; Miethe's last communication from 1920 explains that he had noticed the unusual structure of a crater in Mare Serenitatis with his main instrument while preparing to take pictures of the moon.

In 1921 Miethe directed a science fiction movie, organised a popular press conference and founded a central Testing and Research Institute for Cinema Technology. He died in 1927 at the age of only 65 as a result of a railway accident,[8] his rooftop observatory became isolated until it is moved to Potsdam in 1932 thanks to the activity of Erwin Freundlich of the Einstein Institute for Solar Physics.

Cuno Hoffmeister from the Sonneberg observatory, reported in a foreword written in 1967 that he had received a Zeiss triplet 170/1200 mm from Miethe's holdings in Charlottenburg through the Notgemeinschaft der Deutschen Wissenschaft in 1922, which enabled him to discover new variable stars. In July 1923—still in the roof observatory at his parent-house in Sonneberg's lower city—he received a new mount from Zeiss Jena that could carry his entire current telescope arsenal. In his 1923 annual report he wrote that the objective of the telescope is a three-part apochromat from Zeiss with 13.5 cm aperture and 202 cm focal length. The axis system also carries a large photographic camera with a Zeiss triplet of 17 cm aperture and 120 cm focal length (Fig. 3.4). In late 1925, Hoffmeister moved into a

---

[7] Miethe (2012).
[8] Seegert (1927).

Fig. 3.4 Miethe's 170/1200-mm-camera with Zeiss-triplet from 1913 given 1924 to Sonneberg by the Notgemeinschaft der Deutschen Wissenschaft, now in the museum Sternwarte Sonneberg. (Photo P. Kroll. Courtesy of P. Kroll)

new building erected by the city administration Sonneberg with the support of the state of Thuringia and the Carl Zeiss Foundation. In early summer, the site was found, a 13,000 m² piece of wasteland near the village Neufang, 635 m above sea level. In November, the 4.5 m diameter dome also supplied by Zeiss was installed and in December the main instrument was brought up from the old observatory to the new one, for which the Zeiss Foundation had provided the mechanics. The "building was handed over on 28 December." Right at the beginning of the New Year, he began observations, "which in the following years developed into the Sonneberg field plan"[9] which should basically allow statistics of variable stars including their spatial distribution. On selected fields, all variable stars of any kind up to about the 15th magnitude should be searched for as far as possible. To this end, all discoverers, including amateurs, should observe the variables they have found until at least the type and the elements of the light curve are known. "The discovery of new variables without providing light curves is almost worthless if this rule is not followed."[10]

Now Hoffmeister was the director of a professionally equipped one-man observatory without personnel (Fig. 3.5). The city and the state were unable to pay their chief astronomer a salary, no matter how small. During the economic crisis, the Notgemeinschaft temporarily helped with a performance grant of 250 RM per month, formally for his dissertation. The income from journal articles, the thunderstorm service and the sunshine registrations for the Prussian Meteorological Institute (from 1922) or the

---

[9] Richter & Wenzel (1984). The Miethe triplet was used in Sonneberg until 1971.
[10] Hoffmeister (1929).

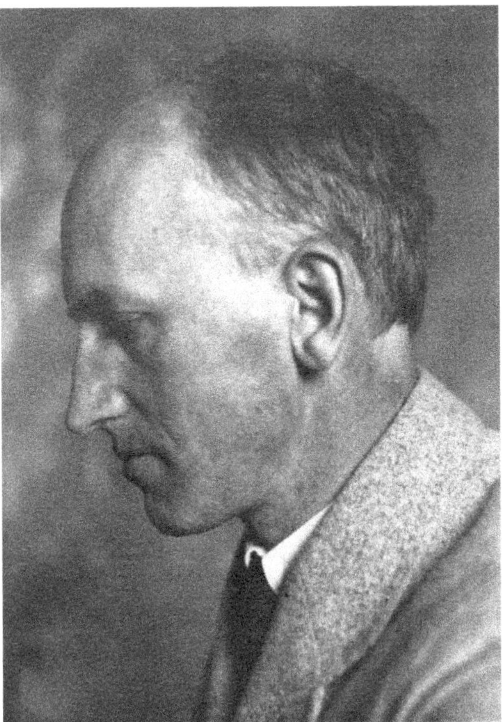

Fig. 3.5   Cuno Hoffmeister (1892–1968) circa 1932. (Courtesy of K. Hoffmeister)

observation of the instruments of the Institute for Earthquake Research to determine the speed of sound in the earth's mantle will hardly have covered his postage costs for the exchange of literature. The solution of the problem came from Berlin-Babelsberg. Director Guthnick of the Berlin Universitätssternwarte wanted to set up photographic monitoring of the northern night sky in co-operation with the observatories in Sonneberg and Bamberg. Using wide-angle telescopes with fast lenses, the entire northern sky was to be photographed zone by zone at least twice a month to locate novae and derive light curves of all the brighter variable stars. Hoffmeister took up the challenge and in September 1928 the first camera came from Babelsberg but soon all four cameras available for monitoring the entire northern sky were in Sonneberg, although the large distance between the observation sites had actually been part of the original concept due to different weather. From January 1929, Hoffmeister's first permanent employee, the optician Rudolf Brandt from Jena, supervised this Babelsberg project (Fig. 3.6). What's more, from 1930, the city administration of Sonneberg handed over the Sternwarte—which was now generously equipped with gas

Fig. 3.6 First 135/240-mm sky observation camera with R. Brandt (1905–1975), here ca. 1929. (Courtesy of Astronomiemuseum Sternwarte Sonneberg)

heating, several buildings and ground—to the Prussian state, which attached it to the observatory in Babelsberg as "Abteilung Sonneberg"[11] Then his Babelsberg colleague Ludwig Biermann[12] later deduced the existence of solar wind[13]—"the discovery of his life" (Kippenhahn)—from Hoffmeister's excessive comet observations.

---

[11] Richter (1992).

[12] 1937–1944. He lived through thethe war in Göttingen.

[13] Biermann (1951a, b).

Finlay-Freundlich and Friends

On 28 May 1932, it is stated that *on the basis of the decree of the Prussian Minister for Science, Art and Education of 16 Nov. the reflector telescope, which had become dispensable at the photochemical laboratory of the Technische Hochschule Berlin, was transferred to the Einstein Institute of the Observatory in Potsdam.* The file lists Erwin Freundlich and Harald v. Klüber as being present but not Director Hans Ludendorff.[14] Shortly before his retirement—under considerable pressure from Einstein—Ludendorff's precursor Gustav Müller hired Erwin Freundlich as an Observer on the Telegraphenberg and allocated the newly established foundation a good building site for the planned Einstein Institute for the experimental development of the Theory of Relativity. Einstein had already written to the young Freundlich in Babelsberg in the summer of 1913,[15] "the astronomers can provide an invaluable service to theoretical physics in the next [solar eclipse] year." After his calculation of the deflection of light at the sun had actually been confirmed by British eclipse expeditions in 1919, everything now depended on the measurement of the predicted redshift—only about 2×10^{-6} of the wavelength—of the solar spectral lines or, as formulated by Karl Schwarzschild, a Doppler effect of the light source of only 635 m/s.[16] If this shift cannot be proven, as it has been so far, the new theory collapses, warned an English physicist. Schwarzschild had already set up the spectroheliograph for precise wavelength measurement of the nitrogen line 3883 Å on the roof of the Beamtenhaus near the later Einstein Tower in 1913, shortly after his arrival as the director in Potsdam.[17] The heliostat reflected the light onto a lens with an aperture of 20 cm and a focal length of 3 m. Possibly parts of the old heliograph from Vogel and Lohse were reused. Schwarzschild spoke of a camera lens with 16 cm and a focal length of 4 m, exactly the data of the original heliograph.

Schwarzschild had found too low redshifts of around 200 m/s on average, lower values for weaker lines and higher values for stronger lines, which even decreased outside the centre of the sun (Fig. 3.7). In 1906 he had introduced his Göttingen paper with "The solar surface shows us changing states and stormy changes in granulation, sunspots and prominences." After the inclusion of the Doppler effect for the descending molecules of the solar gases of

---

[14] Hans Ludendorff (1873–1941), director of the AOP 1921–1938 at the suggestion of G. Müller. Einstein and Nernst had favoured Max v. Laue.

[15] Erwin Freundlich started his work at the Sternwarte Berlin-Babelsberg already in 1910.

[16] Vogel's velocity measurements from 1888 for Arcturus was 7600 ± 600 m/s.

[17] November 1909, against the opposition of Auwers (Dick, 2000), annual salary 11,500 M with free accommodation.

Fig. 3.7 Left to right: Karl Schwarzschild (1873–1916); Gustav Eberhard (1867–1940). (Courtesy of Förderverein Großer Refraktor Potsdam)

around 200 m/s, he even concluded that the smallness of the remnants would prove that the "Einstein effect does not exist."[18] Of course, he also considered the difference between the radial velocity components of the edge and centre leading to Doppler effects of varying strength. Einstein had no problem presenting Schwarzschild's negative result on 5 November 1914 at the meeting of the Academy personally and without resentment. He was pleased, however, that the director of the famous Observatory—who was on the way to the Russian front at this moment—had taken his ideas seriously.

When Schwarzschild had arrived in Potsdam, he replaced the hard-working Hartmann by the 36-year-old Hertzsprung. Hartmann had established with spectra of the binary system δ Ori, obtained soon after the inauguration of the Great Refractor in the winters of 1901 and 1902, that the faint and sharp calcium line K with $\lambda 3934$ does not follow the stellar orbital motions (Fig. 3.8). His conclusion, "between the Sun and δ Orionis there is a cloud that produces this absorption," meant nothing less than the discovery of interstellar matter.[19]

On the other hand, Ejnar Hertzsprung had impressed Schwarzschild with the clever argumentation that the star Capella was actually far too bright for

---

[18] Schwarzschild (1906, 1914).
[19] Hartmann (1904).

Fig. 3.8 One of Hartmann's original spectra (between the artificial lines of an electric arc) of the Orion belt star δ Ori from February 1902. The orbital period of the multiple system determined by Hartmann is 5.7 days, the marked weak calcium line K ($\lambda = 3934$ Å) is at rest despite the orbital motion. (Courtesy of Förderverein Großer Refraktor Potsdam)

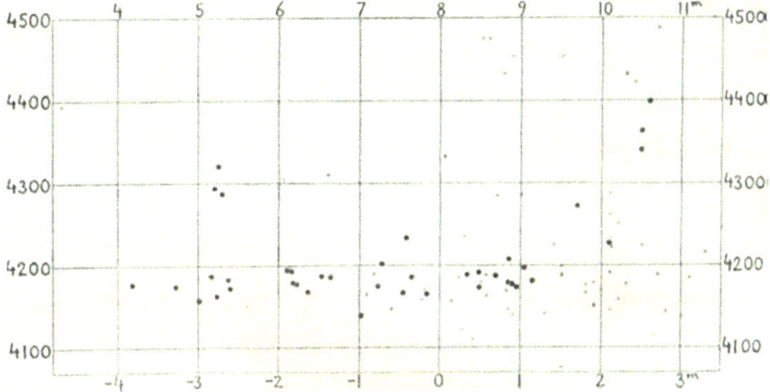

Fig. 3.9 First colour-brightness diagram ever published by Hertzsprung for the open star cluster of the Hyades. Because of the constant distance, the apparent brightness m (abscissa) can be used instead of the absolute brightness. Ordinate: effective wavelength as colour equivalent (Hertzsprung, 1911)

its large distance, meaning that Capella must be of great absolute brightness, although it would be of the same spectral type as the sun. Hence, the two stars cannot be equally bright. Such ideas, combined with his high accuracy, had transformed Hertzsprung with Schwarzschild's help from an amateur astronomer in Copenhagen to a professor of astronomy in Göttingen and Potsdam in short time, without ever having operated a telescope before. In Potsdam he discovered the variability of the northern polar star Polaris which had long been used as a reference star, in 1910/1911 with the 150-mm Zeiss triplet. The colour-brightness diagrams of the Pleiades and Hyades, which were also created at this time, is however based on spectra obtained with the Steinheil astrograph at the Urania Observatory in Copenhagen (Fig. 3.9).

Schwarzschild also began a new round of optical improvements to the Great Refractor lenses. During his short directorship at Potsdam the most notable instrumental improvement was the correction of the two objectives of the Great Refractor.[20] Schwarzschild had commissioned in the fall of 1912 the already famous optician Bernhard Schmidt to correct the 50-cm-lens in his studio in Mittweida within 3 months under the motto "lenses are artwork" against the objections of the Steinheil company. The undersigned is convinced, Schwarzschild wrote to his minister, that Mr. Schmidt is the greater artist. The minister accepted Schwarzschild's opinion and wrote in May 1912 in response to a protesting petition from Steinheil, "I am unable to comply with the request to give you the 50-cm-objective of the Great Refractor of the Astrophysical Observatory in Potsdam for further correction." It would be planned to hand over the correction of this objective to the optician Schmidt in Mittweida. "I am prepared, however, to reserve the decision on the subsequent correction of the 80-cm-lens until the correction of the 50-cm-objective." Schwarzschild's evaluation in July 1913 was that now the 50-cm-lens has been transformed from "a poor lens into a good one, and after covering a 5 cm wide edge zone into an excellent one." A similar treatment of the 80-cm-lens failed due to the desperate resistance not only of the Steinheil company.[21] The cured 50-cm-objective was used by Hertzsprung in the war years 1914–1918 for the search for multiple systems by the photographic examination of 126 visual binaries on 408 plates. Through the clever use of a grating spectrograph, the accuracy of the astrometric location of the binaries was increased tenfold compared to visual observation. "The fact that I have not yet found new pairs with certainty is not discouraging given the short time available."[22]

The Stockholm Nobel Prize Committee had shifted honouring Einstein for his relativity theory from year to year due to a lack of empirical confirmation, despite of very prominent nominations.[23] In July 1918, Sommerfeld warned a friend: I would like to draw your attention … to the redshift of the spectral lines. So far there has been no hint of it. Schwarzschild has not found it, nor have the most careful new American measurements on Mt. Wilson.[24] Einstein hastily wrote to the responsible minister at the beginning of December 1919 that Freundlich in Potsdam was the only German astronomer (apart from Schwarzschild) who supplied outstanding activities to the field (Fig. 3.10). "It

---

[20] Hertzsprung (1917).
[21] Müürsepp (1982).
[22] Hertzsprung (1920).
[23] Pais (1986).
[24] Hentschel (1992), p. 49.

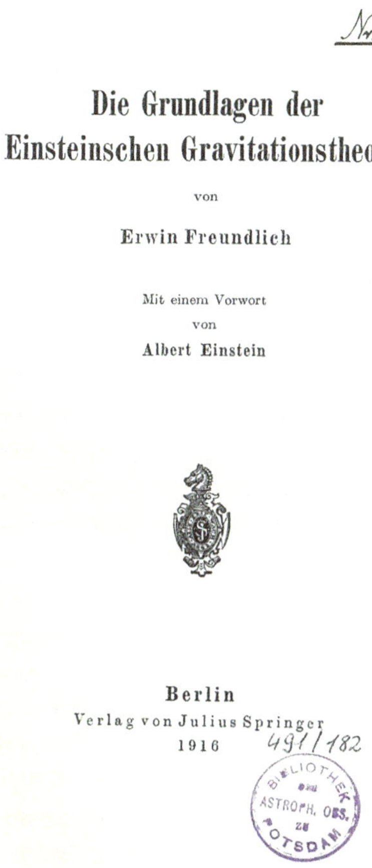

Die Grundlagen der
Einsteinschen Gravitationstheorie

von

Erwin Freundlich

Mit einem Vorwort
von
Albert Einstein

Berlin
Verlag von Julius Springer
1916

Fig. 3.10 Erwin Freundlich (1885–1964) completed his doctorate in Göttingen in 1910 under Felix Klein on "Analytic functions with an arbitrarily prescribed infinite domain" and wrote a textbook on the new theory of gravitation as early as 1916. Einstein to Sommerfeld: "He was the first astronomer to realise the significance of the general theory of relativity"

would be a great support if this astronomer was soon given an observer position at the Potsdam institute with the task of working on testing the general theory of relativity."[25] The long road that Einstein walked to fulfil his Nobel

---

[25] Marginal note on this letter: "The suggestion to support Dr. Freundlich by the Astrophys[ical] Obs[ervatory] came from me after the negative behaviour of G[eheim] R[at] Struve made it impossible for him to remain at the Babelsberg Observatory. … Kr[üss]" (Kirsten & Treder, 1979).

Prize plans[26] had ultimately led to a comeback of solar research at the Telegraphenberg, including the spectacular new building, despite Schwarzschild's negative results which Einstein was aware of and which did not really speak of the success of a new experiment.

The appeal for the Albert Einstein Donation, which was needed to finance the construction of a new tower spectrograph[27] to measure the relativistic line shift, was most probably written in December 1919 by the highly talented Freundlich himself: "Albert Einstein's research on the theory of general relativity signalled a turning point in the development of the natural sciences. ... The experimental testing of its observable conclusions must go hand in hand with the further development of the theory. For the time being, only astronomy seems to be able to probe this work. ... These funds are intended to provide the Astrophysikalisches Observatorium in Potsdam the instruments needed to work successfully on this problem. Approximately 500,000 M are required."[28] The letter carried the handwritten note "Please have my signature at your disposal. The timing is certainly favourable for the campaign! Yours sincerely, W. Nernst." Freundlich later named as 1.19 Mio Marks as the result of his action. The construction of the tower had been going on since the summer of 1920—without official permission—partly because of the actual money inflation. Einstein gave Freundlich on 24 April full procuration "for all matters relating to the construction of the tower spectrograph to be built on the site of the astrophysical institute in Potsdam."[29] Among the numerous donors, the director of the Badische Anilin und Soda Fabrik and later Nobel Prize winner Carl Bosch stood out whose commitment to the Telegraphenberg and the theory of relativity seems to have been inexhaustible.

During the construction period, the young H. Kienle was in Potsdam on a scholarship from the Bavarian Academy of Sciences. The tower building was almost complete in August 1921. In Potsdam's Stadtschloss, Freundlich showed the construction plans of his project and details of the spectral apparatus to the 140 members of the Astronomische Gesellschaft during their first meeting since 1914. Einstein was present to clarify his predictions about the gravitational redshift in the solar spectrum. The confirmation of his theory of

---

[26] Einstein had dedicated the expected prize money to his first wife Mileva. Draft of the divorce agreement (1918, divorce Feb 1919): You may freely dispose of the interests. The capital would be deposited in Switzerland and kept safe for the children.

[27] The vertical construction was intended to reduce the disturbing air turbulence and to enable an increased focal length (Finlay-Freundlich, 1969).

[28] Kirsten & Treder (1979), list of donations see Eggers (1995).

[29] Hentschel (1995). Instrument developed and built by Carl Zeiss Jena including a donation of 300,000 M.

Fig. 3.11 Portrait of Albert Einstein (just awarded by the Nobel Prize) created in 1924 by Kurt Isenstein. It was removed from its podium in front of the tower building and kept in a laboratory between 1933 and 1945. (Photo M. Bautsch. Courtesy of Kunsthal Kongegaarden)

light deflection at the sun by the English expeditions had been published first by the Berliner Illustrirte Zeitung on 14 December 1919 leading to Einstein's world fame overnight (Fig. 3.11). Nevertheless, a newly appointed commission[30] was planned to prepare another solar eclipse expedition to India for September 1922 and Einstein, Freundlich and Nernst surprisingly argued in favour of a perihelion commission in order to use the more numerous and more precise orbital data of the inner planets "for a tighter control of the theory, possibly also on the perihelion motion of other planets." More than a hundred new applicants were accepted as members of the Astronomische Gesellschaft including Einstein and Nernst, 33 foreigners and two women. The closing party with 200 participants took place in the dome of the Great

---

[30] Einstein, Freundlich, Kapteyn, Ludendorff, Schorr, Voute.

Refractor after a visit to the nearby construction site of the solar tower.[31] On 9 November of the following year, Einstein, who seemed overactive at the Potsdam conference, learned that he had been awarded the Nobel Prize in 1921 and was thus only rarely seen later on the Telegraphenberg. His name no longer appears in the lists of participants at the next conferences of the Astronomische Gesellschaft in Leipzig (1924), Copenhagen (1926), Heidelberg (1928) and Budapest (1930).

In June 1932, the *reflector telescope building on the grounds of the Observatory in Potsdam* is officially taken over *by the director of the Einstein Tower*. The 6000 M for the transfer from Berlin to Potsdam were again paid by Bosch. The transfer of the night-astronomy telescope from Charlottenburg to the Telegraphenberg as part of the solar institute took place in the middle of a bitter debate between Ludendorff and Freundlich over the future status of the solar institute as an independent or a dependent part of the AOP. According to the official schedule of the foundation, the institute should have become property of the Prussian state on 1 January 1932, which would then of course have added it to the Observatory. Shortly before this date, the members of the Foundation Board[32] received a note from the Ministry that one of the founders had expressed the wish to "leave the Institute independent as a state institute in honour of its history and not, as would be obvious, to incorporate it into the Astrophysical Observatory. In view of the further development of the tower, I am considering accommodating this wish." An arbitrary decision, the real "institute" is only a solar spectrograph with too few offices and laboratories and without an own building for workspaces. Through the intervention of supporters such as Albert Einstein and Max v. Laue, Erwin Freundlich had won the battle. In April 1931, he thanked Einstein who had operated so intensively "to save my institute," despite the serious differences between the two. He continued that "it wouldn't be easy for me to leave what I've built up here and which would probability be ruined if I left."[33] He mentioned a possible appointment from the Oxford observatory to strengthen his position in Potsdam. At this time, however, it was no longer clear whether Freundlich basically wanted to confirm or reject Einstein's theory. Einstein had invited Laue to a board meeting in Potsdam by writing that "it's going to heat up. One can't live on logic alone; one needs something for the black heart."[34] The installation of the Miethe telescope also indicated a planned departure from the

---

[31] Birck & Pahlen (1921).
[32] Bosch, Einstein (Chairman), Franck, Freundlich, Ludendorff, Müller, Paschen.
[33] Hentschel (1992). Freundlich had applied for a professorship in Istanbul starting 1 November 1933.
[34] Hentschel (1992), Hermann (1996).

original programme of the solar research. Einstein was certainly prepared for anything, as he had already labelled Freundlich a "greyhound" in a letter to Sommerfeld in February 1916, saying that he would not choose him as an intimate friend and that one of his earlier papers had been written in a greyhound-like style. "But he has something to show that is worth its weight in gold, i.e., an enthusiastic dedication to the work; that is a rare quality that he does not share with very many."[35] Much later, Einstein replied to Freundlich's anti-relativity criticisms[36] by writing that he "doesn't move me a bit. If no light deflection, no perihelion motion and no line shift were known at all, the gravitation equations would be convincing because they avoid the inertial system."[37] In response to a simultaneous enquiry from the Academy in East-Berlin, he wrote that "The influence of the Michelson-Morley-experiment on my own ideas was quite indirect."[38] Michelson and his interferometer for measuring the speed of light in Potsdam are not even mentioned in Einstein's early papers.

In July 1928 Freundlich had to suffer a negative comment on his application for a new building to his institute again designed by Erich Mendelsohn: *The project has been worked on by a* _private architect_ *without any involvement of the Bauamt. It is the same architect who designed the Einstein Tower and controlled its construction. Given this experience, does it make sense to entrust this architect with further public buildings?* There were indeed damage reports about the condition of the tower after just a few years. An architect commissioned with one of the latest renovations—without a microphone—did not shy away from the judgement "actually a construction sin." Nevertheless, Mendelsohn was already a seasoned architect (Fig. 3.12). The following year, the final rejection came: *The application for the inclusion of funds for an extension to the Einstein Institute at the Astrophysical Observatory in Potsdam in the draft for the next year's state budget has again not been approved.* The activity of the institute, as long as it can maintain its independent status, depended on workspaces for the increasing number of its employees. This shortcoming of the Tower building is not removed until 1933 with the construction of the so-called Bosch barrack (*made of wood with a cardboard roof, without double windows*) from foundation funds (Fig. 3.13). Freundlich, the networked free spirit, was indeed very successful in raising funds for instruments and assistants; all of his numerous applications for financial support to the

---

[35] A. Einstein, Gesammelte Schriften 8, 255.

[36] Born to Einstein 4 May 1952: Yesterday Freundlich was here [...] It really looks as if your formula is not quite right (Hentschel, 1992).

[37] Einstein to Born 12 May 1952, ibid.

[38] Pais (1986).

Fig. 3.12 Left to right: Erich Mendelsohn (1887–1953) by Erich Stumpp (1886–1941); typical architecture in the tower building (staircase). (Photos Wikimedia Commons Markus Bautsch/Shy halatz)

Fig. 3.13 Einstein Tower with Bosch barrack (foreground) and shelter (righthand) for expedition material. (Courtesy of Förderverein Großer Refraktor Potsdam)

Notgemeinschaft der Deutschen Wissenschaft were approved. "The intellectual liberalism that prevailed in his circle in countless discussions in the Einstein Tower at that time and which was determined by Freundlich and included the entire scientific as well as the human-personal sphere, was one of the best schools that a young scientist could enjoy."[39]

Freundlich never suffered from a lack of ideas. On 29 June 1931 he applied via Ludendorff to the Prussian Minister Adolf Grimme to obtain a second-hand telescope from the Photochemical Institute of the Berlin-Charlottenburg Technical University at a reduced price. *In the possession of the Photo technical Laboratory in Charlottenburg is a mirror telescope with an aperture of 50 cm, which the late Prof. Miethe had acquired who was interested in many questions of astrophotography. This telescope could be used today for many tasks in astrophysics, but has been unused for years.*[40] He also puts this strong argument alongside allegedly planned investigations into a new silvering process for mirrors, *which one of these gentlemen from IG Farben has found in Ludwigshafen, which will make it possible to take pictures of the sky right into the ultraviolet.* The mirrors in the tower telescope would have to be silver-plated twice a year. *It is also important that we ... have the opportunity to familiarise ourselves with the work with a mirror telescope.* A striking plan for a solar institute; Ludendorff promptly rejected the project: *The investigations planned by Prof. Freundlich ... can be carried out with much smaller means in the laboratory ... In particular, the Einstein Institute section of the observatory has a modern instrument in the tower telescope, but very few results have been obtained with it so far.*[41] Ludendorff's comment was aimed at a Communication from the Einstein Institute from 1930, in which a puzzling edge effect was reported, according to which the desired redshift of the Fraunhofer lines occurs at most at the edge of the sun but never in its centre.[42] At the time, the two opponents lived next door to each other, and certainly met almost daily on the sparsely populated Telegraphenberg. Ludendorff lived in one of the observer houses, Freundlich in a new brick building erected for him. The Grotrian and Hassenstein families lived in the director house, also Eberhard and Pahlen lived on the Telegraphenberg.[43]

Minister Grimme from the League of School Reformers approved the application despite such reservations because it was very reasonable to convert

---

[39] Klüber (1965).

[40] Klüber had already used this instrument to obtain the data for his dissertation (Appendix 3).

[41] Archiv der Berlin-Brandenburgischen Akademie der Wissenschaften (ABBAW), AOP, No. 149; cf. Hentschel (1992), p. 131.

[42] Freundlich (1930).

[43] Brück (2000).

a powerful telescope into an active observatory, but perhaps also because of its history, which dated back to the early days of the Weimar Republic. However, the new mirror telescope could not contribute anything to the solar-research profile of the Institute. Freundlich had the new Miethe-dome (a dome larger than that of the solar tower) positioned at the greatest possible distance from the main building of the observatory as a visible sign of his growing distance from solar physics and/or Einstein. Ludendorff had submitted the building application for the Miethe dome himself on 13 November 1931 and the permit came the next day.

Witch Hunting

"Then came the year 1933, in which a large number of the best scientists that Germany had produced left their country forever, among them Einstein himself, while the majority of those who stayed behind hastened to assure the new regime of their devotion."[44] From 30 January on, the successful scientific landscape of the Weimar Republic was permanently destroyed. Minister Grimme had already lost his position in July 1932 when v. Papen became Reichskanzler. In April 1933 the Einstein Institute was given the new name Institute for Solar Physics, after which it was formally incorporated into the Astrophysical Observatory under Ludendorff.[45] Carl Bosch guessed early on what this would lead to. He wrote to the Ministry on 4 April to point out *that the Institute must continue under its present management under all circumstances, because only the intensive contact with outstanding researchers from abroad, which can only be guaranteed by the currently working persons at the Einstein Tower, will enable the Institute to continue to lead German astrophysics.*[46] On 19 May, the Potsdam district leadership of the NS Organisation reported to the government *that Professor E. Freundlich has been employed as the Principal Observer for the Einstein Tower at the Astrophysical Observatory since 1922. As this gentleman is of Jewish descent, we request further arrangements. Heil Hitler!* The date of receipt of this "request" at the Ministry of Science, Art and National Education is 28 June. According to the instruction of 16 June 1933,

---

[44] Klüber (1965). Almost 6000 professors and assistants had to leave the universities in 1933, around 20% of the teaching staff.

[45] Employees AOP 1933: Ludendorff (director), Münch, Freundlich, Westphal, Hassenstein, Grotrian (Principal Observers, salary A2a), v. d. Pahlen, R. Müller (Observers, salary A2b), v. Klüber (scientific assistant, salary 2b), Wurm, Brück (scientific employee). Source: GStAPK, I. HA Rep. 76, Va Sekt. 1, Tit. V No. 4 Adh.

[46] GStAPK, File I. HA Rep. 76, Vc Sekt. 1 Tit. XI No. 6b, Vol. 10.

directors of the independent scientific institutions such as the observatories in Potsdam had to ensure that all employees would declare that they are "not aware of any circumstances that could justify the assumption that I am descended from non-Arian parents or grandparents." The next day, Freundlich applied for a 1-week vacation. Prof. Vahlen from the Berlin Ministry ordered an expert opinion on Freundlich's personality from the director of the Sternwarte Berlin-Babelsberg. Paul Guthnick started with the statement that he had never passed a positive assessment on Freundlich. Guthnick's opinion had *not changed, as Professor Freundlich had not changed either.* Now, however, there were new arguments: *Professor Freundlich has always given me the impression of a fighter for the strengthening of Jewish influence in German cultural life … From the point of view of Jewish imperialist endeavours, to which Freundlich is obviously very close, one must consider his entire activity, especially the founding of the Einstein Institute, which was essentially his initiative … From the point of view of the preservation of German culture, I must unfortunately characterise him as a person to whom I consider it questionable to admit any decisive influence.*[47]

A list of Employees dated 18 July 1933 and signed by Wilhelm Münch[48] does not contain the relaxing entry "not to be arranged" for Freundlich but his name is even underlined in red (Fig. 3.14). On the same day, the Fachschaftsleiter Obst on behalf of the NS-Beamtenabteilung, Fachschaft Observatorien denounced him as an *anti-national thinking Jew descendant,* quoting some old political phrases by Freundlich from a collection of Münch and Wurm.[49] *The entire conversation revealed a completely hostile and insultingly mocking attitude by Prof. Freundlich towards the national movement,* reported Wurm.[50] This was dangerous for Freundlich but not yet enough especially as at the end of June the physics elite of the Academy of Sciences—almost all Nobel laureats—appealed to Minister Rust: *The undersigned hereby take the liberty of once again drawing the Minister's attention to the urgent request to leave the Principal Observer Professor Freundlich in his position at the Astrophysical Observatory Potsdam, which three of them (Laue, Paschen, Planck) had already made on 29 March 1933.*[51] They, however, underestimated Ludendorff: Nobel Prize winners did not impress him as they *cannot possess expertise in the field of*

---

[47] Guthnick to Vahlen 10 July 1933, ibid.

[48] Wilhelm Münch (1879–1969), retired at 1944.

[49] Appendix 1. "Reichskanzler Hitler was described by Freundlich as a 'sectarian', Minister Göring as a 'morphinist'"(Wurm, 10 July 1933); "In a conversation I had with Mr. Freundlich … he described himself as a Bolshevik" (Münch, 11 July 1933).

[50] GStAPK, I. HA Rep. 76, Vc Sekt. 1, Tit. XI, Part II No. 6b, Vol. 10.138.

[51] GStAPK, I. HA Rep. 76, Vc Sekt. 1, Tit. XI, Part II No. 6b, Vol. 10. Appendix 2.

Fig. 3.14 Left to right: Brück, Hassenstein, R. Müller (son of former AOP director Gustav Müller. Rolf Müller was also expert of the ancient history of astronomy, e.g. the orientation of prehistoric sites ("Sun, moon and stars over the realm of the Incas")), unknown, Ludendorff, Grotrian, Münch. (Courtesy of P.M. Bruck)

astrophysics and astronomy, [and did] not consider it necessary to ask me as the director of the observatory about my opinion (Fig. 3.15). Early in October he started a written circular: As I have been informed, the Hitler salute is not yet common practice in the Institute of Solar Physics. I make it everyone's duty … to use this greeting in accordance with the existing regulations,[52] including in contacts with Jews. I shall immediately report any offences of which I become aware.[53] On 7 October, the minister wrote to Freundlich: *I have been informed that the Hitler salute … is not being used in your section. You must immediately inform all employees in your section of the decree and ask whether anyone is*

---

[52] Ministerial decree no. 1712 of 22 July 1933.

[53] Handwritten note: "Transmitted to Prof. Vahlen. H. Ludendorff". ABBAW, AOP, no. 149, cf. Hentschel (1992), pp. 147–150.

Fig. 3.15 Hans Ludendorff was an outstanding expert of "Astronomy, the calendar and the chronology of the Mayan civilisation." (Courtesy of K.-E. Ludendorff)

unwilling to comply with it. The success is to be reported here with your own statement.

The "information" of 5 October 1933 came from Ernst Kohlschütter, Director of the Geodetic Institute at Telegraphenberg, who passed on the statement of an administrative employee. He added that *two officials of the Geodetic Institute, who had initially refused to use this salute, would have been dismissed from their jobs if they had not agreed to use the salute at the last minute.* Freundlich would never give up at the last minute, Ludendorff must have thought, and guessed how to get rid of him quickly. In fact, instead of his signature—provided by the other scientists without comment—Finlay-Freundlich[54] asked *to inform me from which side such a report is supposed to have been received.* The signatures of Brück and Klüber will be missing but Ludendorff will have considered this as being unimportant, neither of them mentions this affair in their autobiographies. Ludendorff responded coolly that he did not owe anybody at the Observatory an account of his actions as director. "The complaints procedure is open to you. I would also like to inform you that the day before yesterday the director of the Geodetic Institute received a complaint against you for not honouring the Hitler salute to a person from his institute. At the request of Prof. Vahlen I handed this complaint over to him yesterday during a visit to the Ministry."[55]

Everything now went quickly. Freundlich replied to the minister's enquiry on 20 October "that all those working in my institute have been informed of the decree on the Hitler salute for weeks and that no one has objected to its observance"—and left for Istanbul on 27 October without waiting for an answer to his petition of 4 October on a two-years leave of absence and without waiting for the end of the campaign directed against him. Immediately, on 28 October, Ludendorff reported to the Ministry that the "*Principal Observer Prof. Freundlich left his post yesterday and went abroad without requesting vacation from me and without informing me of his departure. I have instructed the Geodetic Institute's cash office to suspend salary payments to Prof. Freundlich for the time being*" and added on 28 November that Freundlich had gone abroad (Turkey), taking his furniture with him and that the keys to the empty house had been handed over to a confidant (student).[56] Ludendorff had succeeded, Freundlich was finally taken off the payroll in November, as he left for Turkey at the end of October "*without waiting for the decision on the application for leave he had provided.*" The promised pension will also never be transferred to Istanbul. "*He is not to be informed of this. Should he assert any claims there, it must first report to me,*" ordered the new Ministerial Director Vahlen. Ludendorff was put in charge of the Einstein Tower, and Freundlich received a copy of the corresponding decree. The latter was appointed to the Chair of Meteorology and Astronomy at the newly founded Istanbul Üniversitesi already on 6 July 1933.[57] His application for a 2-year leave of absence as a German official was never answered.[58] Kienle promoted Grotrian

---

[54] Freundlich with a Jewish grandmother on his father's side, had adopted his mother's maiden name ("Finlay") at an early age but used it in scientific publications for the first time in 1927 (and later irregularly).

[55] Pariser Tageblatt, 25 March 1934, facsimile in Hentschel (1992).

[56] GStAPK, I. HA Rep. 76, Vc Sekt. 1, Tit. XI, Part II No. 6b, Vol. 10.

[57] A third of the newly appointed professors came from Germany.

[58] Hentschel (1992).

as Freundlich's successor but on 1 March 1935 Paul ten Bruggencate from Greifswald obtained this position. Ludendorff had intensively supported the talented nuclear physicist Hermann Schüler, but Bruggencate had joined the SA in November 1933.

The newly installed Miethe reflector exceeded all expectations, only the dome building formed a problem. In January 1935, a storm shifted the dome as a whole, and the roof had to be repaired the following year, including the renewal of the outer coating. v. Klüber carried out the fine adjustments for the commissioning of the telescope in a short time and a new electrical fine movement came from the Askania Werke in Friedenau. From autumn 1931 onwards, regular work was carried out in the Miethe dome. The optical quality of the 50-cm-mirror has proven to be excellent. In Ludwigshafen at IG Farben, it received a robust aluminium coating, which, however, had to be carefully protected against condensation. The photographic observations were regularly listed in the annual reports from 1932 onwards, after which Klüber, Becker and Pahlen were persistent main observers of variable stars and open star clusters with the instrument. A publication by W. Krug[59] presented classical photometry in two spectral ranges of the young globular cluster M71, which is considered as one of the most interesting globular clusters today. The necessary images were taken at the 50-cm Goerz-reflector of the Potsdam Observatory. The observers included the results in their textbooks.[60]

After the war, the Goertz reflector, together with other instruments, ended up on a Soviet reparation list; it has been lost forever together with the photo refractor, the Zeiss-triplet and the 40-cm-mirror of the east dome as well as the spectroheliograph system and the only mechanical calculating machine. The list of instruments and apparatus handed over to the Soviet Union consists of 31 entries, signed by Kienle on 21 June 1945.[61]

On 30 January 1941, Kienle as Ludendorff's successor applied for a new construction project to the district president of Potsdam … *first of all to extend the main building in such a way that the available first-floor space is utilised. The Reichsminister for Science and Education will try to include the finances in the 1941 budget.* The plan was already approved on 2 April 1941 from a technical point of view. This success had made Kienle overconfident. As early as 5 April 1941—v. Klüber had just begun to upgrade the tower spectrograph for magnetic field measurements—the archives contain a note "Concerning the conversion and dismantling of the former 'Einstein Tower'" to the same district

---

[59] Krug (1937).

[60] E. v.d. Pahlen: Lehrbuch der Stellarstatistik (1937); W. Becker: Sterne und Sternsysteme (1942).

[61] Kühn (2012).

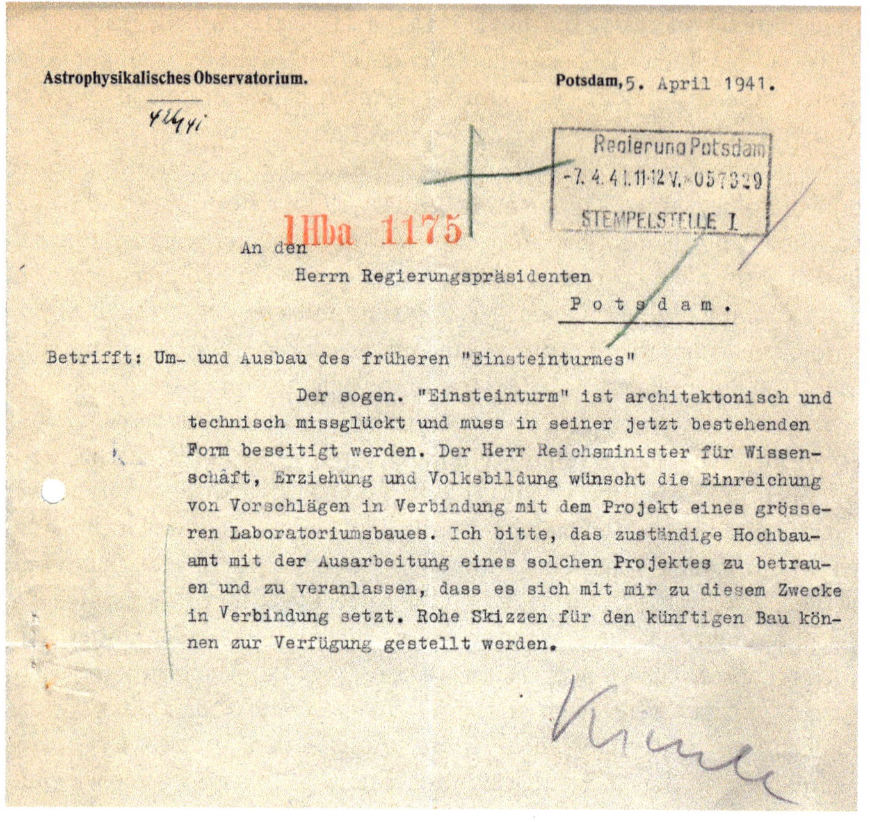

Fig. 3.16 Proposal by Kienle to replace the Einstein Tower with a larger laboratory building. (Courtesy of B. Eggers)

president that *the so-called Einstein Tower failed in technical and architectonical respect* and *must be removed* (Fig. 3.16).[62] The Tower telescope—named here with the old name (!)—only survived this letter because of its meaning for the allegedly war-relevant radio wave research, it remained the only active instrument of the AOP during the WWII. Perhaps v. Klüber had carried out his promising Zeeman experiments in secret or Kienle had underestimated their importance.

Already previously Kienle had made a name for himself as an architectural expert. On 13 October 1939 he informed the President that he needs a garage for his company car and that a *room freed up by the relocation of the kitchen in the basement* would be the most suitable option to house the car. The repeated exchange of letters finally ended in March because *the planned action not be*

---

[62] BLHA Potsdam-Golm.

described as essential to the war success. On the other hand, in 1941 Kienle reported that "at the invitation of the German Scientific Institute in Copenhagen, the undersigned, together with Prof. Heisenberg, Dr. v. Weizsäcker and Dr. Biermann took part in an astronomical-physical meeting, which provided a welcome opportunity for fruitful discussions with Danish colleagues." That meeting with the unknown but legendary conversation between Bohr and his former student Werner Heisenberg on garden paths and without witnesses has entered the collective consciousness of the physics community. There were no joint talks at all, the Nordita Institute in Copenhagen informs today: one nation per table at lunch. In 1943, Kienle and Biermann received the Nicolaus Copernicus Prize in Krakow on the 400th anniversary of the great scientist's birth.

The Solar Magnetic Fields

Since 1941 v. Klüber began to reorient the research of the tower telescope towards the solar magnetic fields; magnetic field measurements in spots and their surroundings are already listed in the annual observatory's report for 1942. Born in Potsdam, Harald v. Klüber as the only descendant of an old family of diplomats and officers (Fig. 3.17). He completed his doctorate in Berlin in 1924 with an observational project carried out at the Miethe Observatory in Charlottenburg. Since 1923, he worked at the Einstein Institute and at the Astrophysical Observatory, holding the position of Observer and professor from 1940 and Principal Observer from 1946. By 1941, Klüber had finally abandoned the task of measuring the gravitational redshift, which was apparently impossible with the tower spectrograph, and had upgraded the instruments to measure solar magnetic fields, until then the unique capability of the Mt. Wilson Observatory. The Potsdam astronomers' focus on the investigation of cosmic magnetic field can be traced back to him alone. He alone organised the turning point in the history of the Einstein Tower and the history of the AOP, towards the study of the magnetic sun. To measure the Zeeman splitting of the 6302 Å iron line by the magnetic field of sunspots, wavelength differences of Å/1000 had to be resolved.[63] The optics of the Einstein Tower telescope and its laboratory indeed made it possible to carry out such investigations.[64] Klüber used a trick to separate the two minimally shifted line components by exploiting their different polarisation, so

---

[63] Similar to a measurement of the protometer to a 0.001 mm tolerance.
[64] Bruggencate & Klüber (1947).

Fig. 3.17 Harald von Klüber (1901–1978). (Sterne und Weltraum 18, 40 (1979). Courtesy of Sterne und Weltraum)

that "in the two superimposed spectra, the long-wave side component of the longitudinal Zeeman doublet is cancelled out in the upper spectrum and the short-wave side component in the lower spectrum."[65] The method was very effective, a spot photo only required about 20 s of exposure time (Fig. 3.18). The weak large-scale magnetic field of the sun had already been successfully observed in Potsdam during the 1944 spot minimum with a theoretical accuracy of ±10 Gauss, although the published values were still too high. The Einstein Tower had nevertheless found its successful future.

A waste of magnetic field measurements was accumulated in this way until the 1950s without coming closer to the explanation of the solar cycle. Finally, Klüber describes the fortunate composition of the Einstein Tower team with Freundlich, v. d. Pahlen, Grotrian and the mechanic Erich Strohbusch. He could not have found a better place in Germany at this time of rapidly developing research.[66] He would never have been forced into the military during the war, but like many of his colleagues he had to carry out extensive

---

[65] Brunnckow & Grotrian (1949).

[66] Appendix 3.

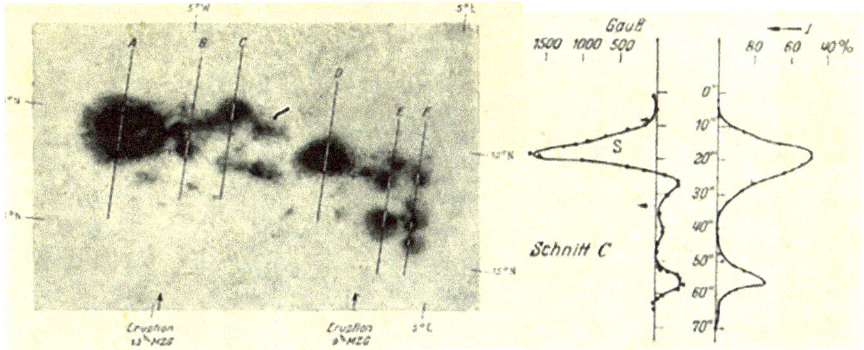

Fig. 3.18 Left to right: Measurement of a group of spots with positions of the spectrograph slit; field strength and brightness at section C (Bruggencate & Klüber, 1947)

calculations on astronomical navigation. He seldom left Potsdam during this period. All of the observatory's telescopes that were not used for solar observations were decommissioned due to the war with the exception that in 1940 the 80-cm-objective of the Great Refractor was revised by Carl-Zeiss. Grotrian, as the military chief of a radio wave research detachment, ordered the soldier Künzel home to support Klüber's sunspot observations in Potsdam. On the other hand, Klüber's knowledge of ionospheric physics was so detailed after the end of the war,[67] that he was certainly involved in the military-motivated activities of Grotrian and Kiepenheuer.

The 1929 Eclipse Expedition

In August 1914 the not yet 30 years old Freundlich was the first astronomer in the world who wanted to demonstrate the gravitational deflection of the starlight during a solar eclipse, as predicted by Einstein with the (wrong) Newtonian value. His expedition to the Crimea was largely privately funded (Krupp Foundation, Academy of Sciences) and did not belong to any official institution. The war destroyed his plans, he was forced to leave the location, interned in Odessa and later exchanged in a prisoner swap. Moreover, the spectacular eclipse of 1919 was tragically out of his reach for political reasons. Others had realised his ideas derived from his theoretical background. He would not find peace with Einstein's predictions until the 1950s. His experiments in 1922 and 1926 remained without measurable success,[68] but the

---

[67] Klüber (1947).
[68] Grotrian (1952).

Fig. 3.19 Double horizontal camera with heliostat and deflection mirrors, focal length 8.6 m, assembled on the Telegraphenberg. In the background the Beamtenwohnhaus with spectroheliograph. (Freundlich, 1930)

concept of the giant Potsdam double horizontal camera was developed step by step, with the first tube aimed directly at the eclipsed sun, while the second one was pointed at a distant star field which did not experience any light deflection for comparison. Both cameras had a focal length of 8.6 m and received their light from one and the same Zeiss heliostat (Fig. 3.19). In 1929, from the total eclipse under a clear sky, four images of the solar neighbourhood and three of the comparison field were taken, each on large 45 × 45 cm plates. A fourth cassette has been lost. For comparison, identical night photos of the star fields at the same observation site were required; Klüber completed this task exactly 6 months later, which for the sake of simplicity he spent on Sumatra with extensive local expeditions through Sumatra, through Java and Bali.

v. Klüber was a highly active participant, organiser and leader of solar eclipse expeditions throughout his life. The eclipses of 1923, 1926 and later 1952, 1954, 1955,[69] 1958 and 1959 were mostly used to investigate the spectral properties of the corona. The well-prepared Potsdam expedition to Takengon (North Sumatra) on 9 May 1929 was a particular exception because

---

[69] In Ceylon, together with Mattig and Strohbusch, unsuccessful due to weather conditions.

it was dedicated to a seemingly solved problem. Ten years earlier, two British expeditions to Brazil and to an island of Spanish Guinea had measured the deflection of light from stars close to the sun and, with 1.98″ and 1.61″, almost exactly provided the Einstein value of 1.75″. Before his departure Eddington had predicted that "These eclipse expeditions will perhaps prove the weight of light for the first time; or they will prove Einstein's strange theory of non-Euclidean space; or they will produce a result with even more far-reaching consequences—no deflection."[70] The result of the expedition seemed to confirm the strange theory, establishing Einstein's world-wide fame overnight.

In the winter of 1928/29, the Potsdam expedition set off by ship from Antwerp to Sumatra with 15 t of luggage in 70 boxes. With their technical strategy—confirmation or correction of the results announced by Eddington—the Potsdam group was not alone. R. Süring as the head of the meteorological institute informed in the Berliner Lokalanzeiger that an important item on the programme of the German, American and Japanese expeditions is "the investigation of the deflection of light from stars close to the sun as they pass through the solar atmosphere in order to test Einstein's ideas." Under the direction of Erich Strohbusch two large devices including a complete electric power station and radio equipment were brought to the Sumatran jungle for connection to the long-wave antenna in Nauen/Berlin. The classic eclipse instrument was an improved duplicate of the astrograph with a focal length of 3.4 m that Freundlich had lost on his expedition in 1914 (Fig. 3.20). The team Strohbusch/Klüber managed to take 6 images on 3 plates during the 5 min of total eclipse, 3 images of the stars surrounding the sun (up to 60 s exposure time) and 3 images each of a comparison field. Three times the long tube had to be moved in deep darkness from the location of the sun to the comparison field. With the natural excitement of all participants, special care must be taken to minimise any uncertainty, was a typical order of Freundlich.

The weather conditions on the eclipse day must have been nerve-wracking. By 8 a.m., 5 h before the event, there was complete cloud cover so that even the first contact of the moon with the sun remained invisible. However, there was a full clearing before the onset of totality. Shortly before the complete eclipse, "the lighting conditions in the atmosphere are so unusual that one loses all feeling about the sky. The fact that there was clear visibility in the vicinity of the sun was shown by the corona, whose radiant structure stood out unusually clearly. Many bright stars were

---

[70] Pais (1986).

Fig. 3.20   Erwin Freundlich during the solar eclipse expedition 1929 on North Sumatra. Parallactically mounted astrograph with a focal length of 3.4 m. (Courtesy of Förderverein Großer Refraktor Potsdam)

visible."[71] Klüber in his obituary to Freundlich wrote that the extensive reductions of the plates carried out at the Einstein Institute 1930–1933 gave the amount of light deflection with the smallest mean error yet achieved, but about 30% too large (Fig. 3.21). Almost all earlier and later determinations also "seem to indicate a somewhat higher value, and Freundlich pursued the clarification of this phenomenon throughout his life."[72] The Potsdam group had even argued that the globally celebrated values of the British expeditions of 1919, which well-fit the theoretical prediction, were based on wrong reductions. Einstein did not really enjoy the work of his friends from the Einstein Institute an animus that for the rest of his life. To be on the safe side, he emphasised the beauty of his theory: I do not see the main significance of the general theory of relativity in the fact that it has predicted a few tiny observable effects, but rather in the simplicity of its basis and in its consistency.[73] This sounds

---

[71] Freundlich (1930).

[72] Klüber (1965).

[73] Pais (1986).

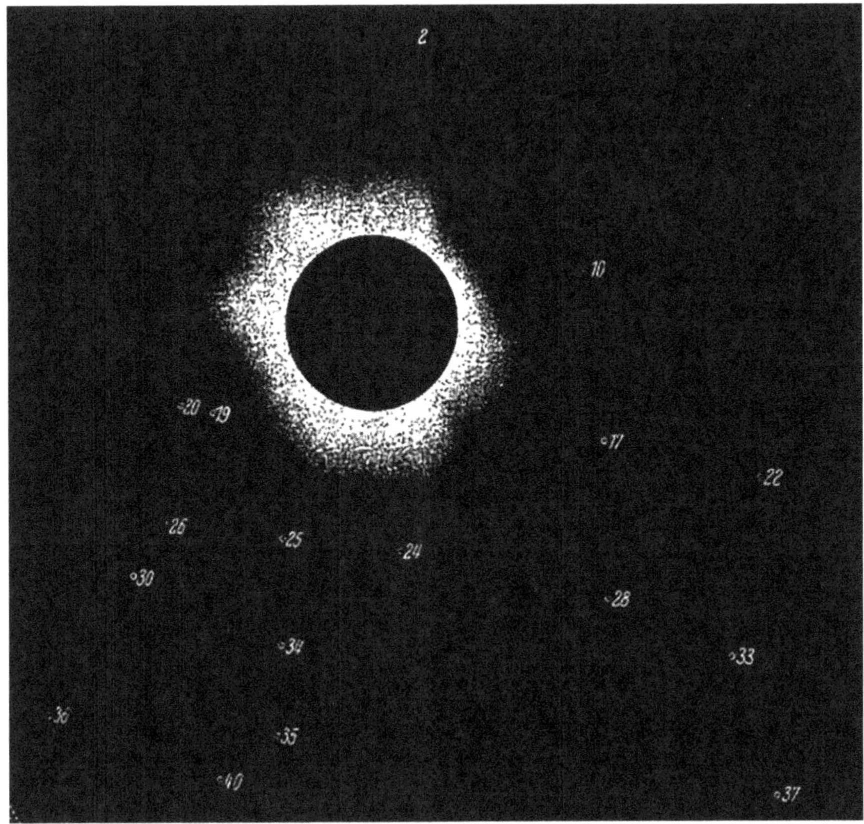

Fig. 3.21   Total solar eclipse of 9 May 1929, one of the four successful Potsdam images. The measurable surrounding stars are marked. (Freundlich, 1931)

surprisingly. Before his Nobel Prize, at the beginning of 1916, in a letter to Schwarzschild stationed in Russia, he wrote on the possible measurement of light deflection on Jupiter: As far as Jupiter is concerned, I recognise that it is a difficult task for astronomers. But in my opinion, the importance of the subject justifies only *one* point of view, and that is: It *has to* work![74] The gravitational redshift is today considered to be confirmed to 5% by solar observations, the deflection of light to only 20%.[75]

After the expedition of 1929 Freundlich formulated: If the plate material which is still being processed leads to the same [Eddington] result, there will no longer be any doubt that the deflection of light predicted by the theory

---

[74] Hentschel (1995).

[75] W. Kundt in: Springer tracts in modern physics 47 (1968).

does exist; but if it had a greater value then it will be the task of the future to develop the theory of relativity in such a way that it is able to explain this phenomenon.[76] Freundlich was certainly a relativist but he wanted to be involved in the development of the final version of relativity theory—not simply to confirm its existing form. After the Sumatra expedition the positions on Einstein had changed which must have amused Ludendorff. In a new paper, he argues that the expedition data, if analysed slightly differently, would lead to a light deflection of 1.90″,[77] which the gentlemen from the Einstein Institute loudly denied. The permanent dispute on Telegraphenberg could not have been presented to the public more precisely. Still in 1952, at the 17th German Physics Conference in Berlin, Freundlich insisted that the Potsdam eclipse result of 1929 was too high compared to Einstein's predictions and he said that it must be expected that the observed phenomena of the propagation of electromagnetic energy (photons) in the gravitation field of the sun will not be fully described by the very formal and abstract solutions. He had expressed similar scepticism about the gravitational redshift in 1930: The manifold attempts in America and England to prove a general shift of the solar lines by 2×10^{-6} times their wavelength had not reached any conclusive result. "Although the existence of the effect is proven with a high degree of probability, the measurement results to date cannot be condensed into a conclusive proof in favour of the theory of relativity."[78]

During the eclipse, Eva und Walter Grotrian unspectacularly used the three Potsdam spectrographs to produce six spectra tangential to the moon, the analysis and interpretation of which, after years of work, will be by far the greatest success of the Potsdam expedition.[79] Already on 1 October 1932 Grotrian became Principal Observer at the AOP. Additionally, the images showed the line spectrum of a bright prominence at the edge of the sun, whose systematic observations he was only able to continue decades later as the director of the Potsdam Observatory.

---

[76] Freundlich et al. (1931).

[77] Ludendorff (1932).

[78] Finlay-Freundlich (1930, 1953).

[79] Mattig (1999).

A Coronagraph for the Einstein Tower

After the departure of Klüber who, at the age of 46, left for ever his lifework on the Telegraphenberg and his villa in Potsdam,[80] Grotrian became head of the Einstein Tower after a 6-year absence due to war. From 1 January 1947, the German Academy of Sciences in Berlin (AdW) took over the institutes on Telegraphenberg alongside the Babelsberg and Sonneberg observatories by order of the Soviet administration and made astonishing investments. For the reopening of the Academy on 1 August 1946, Kienle was asked to give the opening speech "The scales of the cosmos." He pursued and achieved the integration of the astronomical institutions into the Academy from the very beginning.[81] The old project to add a storey to the attic of the main building, was realised in 1948, finally creating seven new offices. In 1950, on the 250th anniversary of the Academy, the bomb-damaged Tower also received a general reconstruction but without the completely destroyed Bosch barrack. In the west dome, the Grubb-refractor with the new lens from Steinheil survived, while in the east dome in 1949 a self-built Schmidt camera was mounted. Professor Grotrian who had never been a party member[82] became Managing Director on 1 October 1950 and Director of the Institute from 1 January 1951, but it was only a tiny command that he took over (Fig. 3.22). In WWI Grotrian was a front-line soldier and flight-observer. In WWII, as a major in the Wehrmacht, he was the military head of a special unit for ionospheric research. Only Wempe had remained from the war period,[83] the observers Daene and Schneller as well as Brunnckow as assistant had joined at the turn of the year 1948/1949. The astronomers Hassenstein, v. Klüber, H. Müller, R. Müller, Münch and Pahlen had left for various reasons. In his first annual report, Grotrian stoically wrote about the former director Kienle that "war-time and post-war times presented him with difficult tasks. The observatory thanks him for his preservation and reconstruction in connection with the integration of the Astrophysical Observatory into the Academy." No word about science. Johann ("Hans") Kienle headed the AOP from 1939 to 30 September 1950, after which he worked in Heidelberg where he was able to

---

[80] On leave 1947, destination Zurich. From 1942/1943 v.d. Pahlen permanently went to a sanatorium in Davos, retiring 1944 in Potsdam. W. Becker took over Pahlen's later professorship in Basel in 1953.

[81] Kienle (1949).

[82] According to Kuiper (1946) non-membership of German astronomers in the NSDAP during the war was exceptional—there were, however, prominent counter-examples at the AOP (Grotrian, Kienle, v. Klüber, Wempe). Ludendorff was nationalist but not a member of NSDAP (Tiemann, 1991).

[83] He worked for Astronomische Nachrichten from 1947 and became Principal Observer on 1949.

Fig. 3.22 Left to right: Kienle (1895–1975); Künzel (1921–2013); Grotrian (1890–1954); unknown. (Courtesy of Förderverein Großer Refraktor Potsdam)

maintain his excellent connections to the central management of the East-Berlin Academy of Sciences over more than one decade.

Shortly before the WWII, Grotrian had made pioneering discoveries about the nature of the Nova Herculis and of the solar corona and was now studying the solar magnetic fields full-time. The magnetic fields of sunspots are the phenomenon in which it is most clearly and conspicuously apparent that electrodynamic forces also play a significant role in the interplay of all the forces that determine the stellar structure, he wrote in his very last review.[84] In an early plan for astronomy in the Soviet Zone for 1945/46, the *state of the sun and the connection between solar and terrestrial phenomena* was formulated as his new research goal, in the very spirit of the old Förster. Later he formulated that *sunspots, corona, prominences and the measurement of temporal changes in magnetic fields are of central meaning.* Grotrian wanted not more than to solve the old puzzle of the 11-year Schwabe cycle by focussing on measurements of the solar magnetic fluxes.

Grotrian also wished to know whether the shape of the sun is influenced by the cyclic magnetic field. He needed an instrument to determine the solar diameter, but even more pressing was the lack of a simple coronagraph to define the edge of the sun and directly track the prominences up to high latitudes.[85] Klüber enthusiastically describes how he was allowed to work with the coronagraph at the Arosa Observatory for almost a year in 1948. At this time, the connection between prominences and the sunspot cycle had already become part of the international observation programs.[86] Grotrian in Potsdam also wanted to be present when the mysterious magnetic corona heating was clarified.

Another motive was radio wave research. In 1946 Klüber had stated that observations of the corona at the eastern limb of the sun indicate those places on the sun a week later which could possibly produce increased ionization of the ionosphere on Earth. "Corona observations are thus particularly important for forecasting shortwave propagation."[87] Even Kienle had wanted a coronagraph for the Telegraphenberg to observe the solar limb.[88] In coronagraphs the bright solar disc in the main focus is covered by a conical diaphragm, as in a solar eclipse, in order to receive only the faint light from the inner corona. Kienle had failed due to political difficulties. The then powerful Kiepenheuer had told him in the fall of 1942 that "all orders for solar physics in the Reich" could only come from him. Due to the low brightness of the corona, the instrumentational scatter light had to be suppressed by carefully processing the lenses. Kiepenheuer had denied the Potsdam team the necessary support.[89]

The Miethe dome had been empty since 1945. Grotrian had worked with E. Strohbusch to build a coronagraph with an aperture of 13 cm and a focal

---

[84] Grotrian (1952).

[85] Waldmeier (1941).

[86] Behr & Siedentopf (1952).

[87] Klüber (1947).

[88] Kummer (1996) with an account of Kienle's clever handling with the utensils of the Crimean Observatory stolen by the Wehrmacht in 1943/44.

[89] Seiler (2007).

length of 190 cm for spectrohelioscope observations of time-varying processes on the sun and was clever enough to take advantage of an internal academy reconstruction programme for the Berlin area. A construction company from Potsdam was commissioned to build a new 5-m Zeiss-dome on the foundation walls of the Miethe building with a photo darkroom and a tiny office (Fig. 3.23). In a letter dated 13 November 1952 to the Academy's Department of Mathematics and Natural Sciences, Grotrian and Brunnckow as the two remaining solar physicists at the observatory described the future work as *Investigation of the magnetic fields of sunspots and evaluation of the observations from 1946 to 1951 with regard to polarity and maximum field strength.* Unfortunately, Brunnckow was unable to continue his work in Potsdam "due to passport difficulties," as the annual report for 1953 laconically states.

It is possible that Grotrian had recognised the limited significance of his magnetic field measurements on sunspots for the explanation of the spot cycle. He wanted to investigate prominences as the other expression of the sun's magnetic activity. Already 100 years earlier Carrington reported an eruption of bright solar spots on 1 September 1859, which lasted less than 5 min and caused a huge magnetic storm on Earth the following night. It was not until the beginning of 1954 that Grotrian's own coronagraph was fully installed in the new dome. It was put into operation the following year,

Fig. 3.23 Miethe house, today A23. The 5-m-dome was removed after 1990 under unclear circumstances. (Courtesy of Förderverein Großer Refraktor Potsdam)

complete with guide tube and recording camera with 6 × 6 cm roll film—too late for Walter Grotrian.

From the very beginning, the guide tube of the new instrument had also been used for visual and photographic solar monitoring (Fig. 3.24). From May 1955, the entire solar limb was photographed to determine the position of prominences. Working fast it was even possible to follow the explosive rise of the prominences. For the International Geophysical Year 1957/58, special funding enabled particularly intensive monitoring of the sun; the coronagraph provided photographic images of the solar limb on 76 days, which were sent into a database at the Fraunhofer Institute in Freiburg. The coronagraph material was never analysed for research, perhaps because Potsdam's geographical location could not keep up with other instruments on high mountains, but mainly because of the huge gap left by Grotrian's death in early 1954. Later, the dome was overgrown by the surrounding trees. In April 1985, the instrument was moved to the east dome of the main building and used for measurements in the chromosphere by attaching filters. It has been at rest there for a long time and observations have been cancelled.

In his diploma thesis, Grotrian's PhD student E.H. Schröter had discussed the almost circular isolines of the magnetic field of a sunspot, measured by rotating the image of the sun across the fixed slit. He received his doctorate from the Humboldt University of Berlin in 1956. He successfully reinterpreted old redshift measurements from the tower spectrograph after re-discussion of the turbulent flow conditions on the sun, which had previously been ignored. After him the often-lamented difference between

Fig. 3.24 Grotrian's coronagraph (guide tube), 13 cm refractor with 190 cm focal length, originally set up in the Miethe dome in 1952, today in the east dome of the Michelson House of the Potsdam Institute for Climate Impact Research

the measured and the expected values can be understood as due to turbulence. "Even if our limited knowledge of convection and turbulence in the solar atmosphere leaves quite a lot of space for … assumptions, the addition of unknown physical processes does not seem to be necessary",[90] the candidate Schröter informed Freundlich in a preliminary report in January 1955. Magnetic fields and turbulence—Klüber, Grotrian and Schröter had fixed the direction that the work of the entire observatory will follow to its end.

---

[90] Schröter (1955, 1957).

4

The Daddy on the Hill

A Summer Piece

In the year 1951, the 61-year-old Walter Grotrian followed his new passion of researching the solar cycle with magnetic-field measurements. The actual idea was to measure the magnetic flux formed by all the sunspots on the surface over days, months and years in order to determine where the magnetic field moves during the spotless minimum that occurs regularly every 11 years. The dissipative power of turbulent convection was not yet known, although one could observe it easily from the decay of isolated sunspots. According to the then prevailing theory of Alfvén the spots on the sun were magnetised vortices that were supposed to travel along the solar dipole field in form of MHD waves from the convective (!) solar core to the surface. "The initial cause of the sunspots is the convection in the solar core."[1] A complicated mechanism in the unobservable core area may ensure an 11-year cycle and zone migration of the spots to the equator. Grotrian's data came from California and from the Einstein Tower. The external measurement results had sent to friends in West-Berlin, where his driver was able to pick them up, undisturbed by controls. As a full professor with a chair at the Humboldt University, he was very well-known, privileged, a full member of the Academy and one of the very first GDR state award winner, at that time still with the claim to be an all-German prize.

---

[1] Alfvén (1950), Appendix 4.

As early as 1921, the young Grotrian can be seen in a famous Berlin photograph right next to Albert Einstein taken in Dahlem on the occasion of the departure to Göttingen of James Franck as the future chair of expermental physics there. There Grotrian had trained under his supervision as an experimental physics assistent. At the end of 1922, he accepted the offer to work in Potsdam as an assistant at the newly established Einstein Institute for Solar Research. There he discovered 1939 that the mysterious red corona line λ6374 was actually a forbidden emission of the highly ionised iron atom Fe X, which had lost almost half of its electrons and which had remained unknown until then because it did not occur in the laboratory.[2] Its excitation requires extreme temperatures of millions of degrees. However, he has never written down this actually unmistakable consequence of his discovery, he avoided theoretical discussions and finally did not dare to give a numerical value, although he was regarded as the leading spectroscopist in Germany after publishing his famous term schemata of the permitted quantum transitions between the energy levels of the electrons in atoms. Just 2 years later Alfvén had announced on the second page of one of his first publications that the sun's atmosphere was "absurdly" hot, much hotter than its surface, which would contradict the simplest laws of physics because heat always flows towards the lower temperature, never the other way round. He argued that the magnetic waves he had recently discovered should be able to heat up the solar outer atmosphere.[3]

Grotrian had not only discussed forbidden lines of the solar corona. W. Münch had begun in mid-December 1934 to produce spectra of the recently discovered Nova Herculis 1934 (DQ Her) with the photographic 80-cm-refractor, which Grotrian was able to analyse with a student. All absorption lines proved to be blue-shifted for expansion speeds of about 300 km/s, hydrogen even for 1000 km/s, whereas the emission lines characteristic of novae and planetary nebulae were mostly not. He concluded that this is an expanding gas nebula from the largely symmetrical contours of the emission lines. As they are broadened by all radial velocity components, he ruled out any noticeable coverage by the central star, so it is obviously very minor. "In this context, however, the forbidden O I lines must be mentioned in particular,"[4] which only occur at very low densities. With the help of the theory of Milne (all ex-novae, as well as the nuclei of planetary nebulae, may be considered as white dwarfs), Grotrian reconstructed the complex light

---

[2] Grotrian (1939) with the same argumentation for λ7892 (Fe XI).

[3] Alfvén (1941). L. Biermann wanted to heat the corona still in 1947 by accretion of interplanetary matter from outside.

[4] Grotrian & Rambauske (1935).

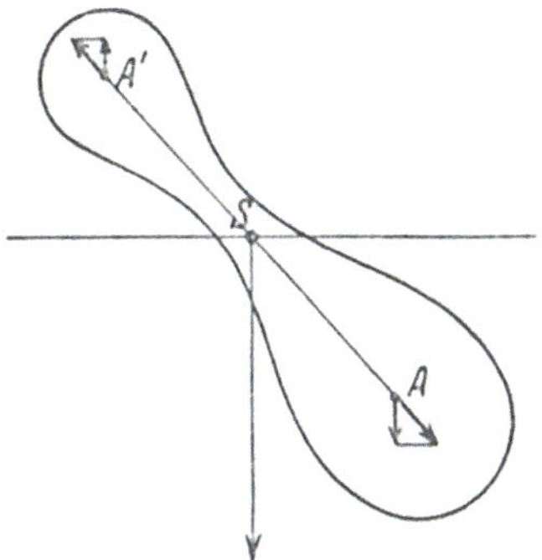
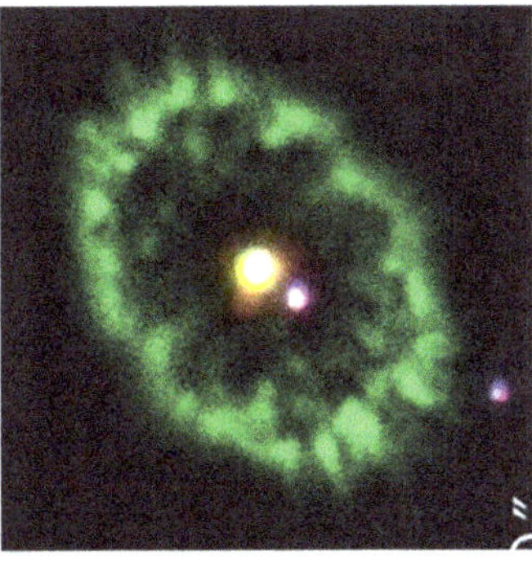

Fig. 4.1 Left to right: schematic representation of the nebular envelope of Nova Herculis 1934 (DQ Her) by Grotrian (1937); a modern false colour image of the expanding gas disk. The line connecting the two stars defines the equatorial axis of symmetry of the nebula. (ApJ 892, 60 (2020). Courtesy of E. Santamaria)

curve of the nova using radiation gas dynamics calculations for the expanding thin envelope.[5] He recognised that the envelope—in contrast to Milne's concept—cannot be spherically symmetrical, but has a "shape that corresponds to a central sector cut out of the sphere around the central star with a certain opening angle" (Fig. 4.1). Even rotation comes into play: If, as a result of rapid rotation, the star "would have the shape of an oblate ellipsoid, whose radiation would preferentially go in two directions exciting the nebular masses in these directions to glow." It can also be seen that in the course of a few days, the nova has grown from the diameter of the sun to about half the diameter of the Earth's orbit. "It may be left to the reader's imagination what will happen if the sun itself entered the nova stage." It would have relaxed him—and not only him—to know that the nova stage belongs to the cataclysmic binary stars which consist of an M-type star and a white dwarf of a distance of only 1.4 solar radius and an eclipsing period of about 5 h. Certainly, only the immanent WWII will prevent the double star character of the Novae from being discovered in Potsdam.

Grotrians current magnetic field calculations were data-intensive; the spots move from day to day; they also decay rapidly and the observed areas and fields are constantly changing. A single mechanical calculator of that time could help with the addition and multiplication. Since last autumn, the young assistants Helga Starke and Ilse Schischkoff had been employed to support the calculations. Starke after finishing school in Thuringia wanted to study astronomy and had asked her widowed father to move to relatives living in Werder

---

[5] Grotrian (1937).

Fig. 4.2 Astrophysical Observatory (circa 1950); from left Direktorhaus (former residence of Vogel, Schwarzschild, Kienle, Grotrian, Jäger), Main building, Great Refractor, Beamtenwohnhaus, Einstein Tower. In the foreground the Photo dome for the sky map refractor. (Courtesy of H. Sperlich née Starke)

near Potsdam. The observatory couldn't have been better for her plans: textbooks everywhere, entertaining colleagues in every room, even the fulfilment of her greatest wish to work with the Great Refractor, which was still damaged at the time, was not without reach (Fig. 4.2). It was a dream with family connection. In the Direktorhaus, right next to the main building of the observatory with the new second floor Mrs. Grotrian regularly invited to lunch (Fig. 4.3). In the evenings, they played music or cards and when it got late, some stayed overnight in one of the many rooms. The food was often from West-Berlin where the driver knew the best places.

In the early morning on Wednesday 29 August 1951, the doorbell at Chausseestraße 169 in Glindow near Werder rang loudly. Two men in leather coats waited in tense attention, a car with running engine stopped in front of the house. Alfred Starke, a finance-officer at the transportation centre of Brandenburg, who lived alone with his daughters, opened the door and the persons were immediately in the hall: a criminal investigation. Your daughter Helga has to go to court in Potsdam to clarify a matter! The young woman was still calm, it could have something to do with the arrests 2 months ago in Werder, she thought, they had heard about it, also from leaflets. She even knew some of the young people from dances and night parties—or maybe

Fig. 4.3 Walter Grotrian (1890–1954) and Eva Grotrian née Merkel (1891–1958) in the Direktorhaus of the AOP. (Courtesy of Förderverein Großer Refraktor Potsdam)

something had gone missing at the observatory. She pulled a light wind-breaker over her summer dress in case it got late in the evening.

She remembered that a week ago she had visited her cousin, who had left Werder for West-Berlin. He had wanted to meet the former leader of a private theatre group. In his flat, they talked about the current arrests in Werder and finally she had received an anti-communist satirical magazine to pass on to Potsdam. Later, over the months, she had been given a handful of closed enve-lopes—which she had never opened—to put in Potsdam post boxes to save postage. This kind of thing was part of everyday life in the four-sector city, other things besides food and magazines were smuggled across the almost open border—it couldn't be that, she assumed. In Potsdam the car turned towards the remains of the destroyed Stadtschloss. It drove into a barrack area, the guarded gate closed behind the car and the 19-year-old was in hands of the Ministry for State Security (MfS) in a tiny dark cell, wearing a summer dress and a thin jacket. After a few days, she was given her belt and shoelaces back and thought she would be released, but the transport stopped after just a few minutes at nearby Lindenstraße 55, the actual prison of the Soviet secret police. The civil Potsdam prosecutor had previously reported the arrests to the Russian military authorities and listed the planned prison sentences of a few years only. As a consequence, the Russian military jurisdiction claimed most of the young people for sentencing including Helga Starke.

On the same day, the Physical Review published an article by scientists from the U.S. Naval Research Laboratory, the technology laboratory of the Navy and Marine Corps in Washington, DC. The head of the electron optics division, Friedman had developed Geiger counters for hard UV light and X-rays years ago, with which the Heaviside layer of ionised oxygen molecules known from radio technology was to be measured at an altitude of 100 km. Because this layer showed clear day-night changes, it was assumed that the X-ray light required for its ionisation came from the sun. Herbert Friedman had heard that more than a hundred German A4 rockets had been moved from Thuringia to the desert of New Mexico, including their documentation and German specialists, Wernher von Braun again produced secret military equipment there. In February 1949, a new altitude record of 400 km was set with a two-stage version of the A4; in total, more than 70 German units were fired over the years. In September 1949, they had launched one of these rockets, in whose empty warhead Friedman's photon counters had been installed as the payload. Kiepenheuer had followed similar plans in 1944 but was not provided with an A4 missile.[6] The rocket showed a stable flight until the burn over, after which it slowly rotated around its own axis and after a few min reached its peak altitude of 150 km, where the instrument part was thrown off. The counters had registered UV light at 70 km and X-ray at 87 km.[7] As the surface of the sun is far too cold to produce such high-energy radiation, it was clear that the corona must be millions of degrees hot, as Grotrian's finding 12 years earlier had suggested. The actual mechanism of this heating is still unknown today. The sun as an X-ray star, a spectacular new chapter in the astronomy book had been opened with participation from Potsdam. "Tell them a good story and you will get all the support," Friedman will later formulate the future world-wide science system.

Missing Miss Starke

Grotrian will have heard this spectacular news very soon as he was also managing director of the German Physical Society during WWII under Carl Ramsauer who lived in West-Berlin. Now having just returned from the conference of the Astronomische Gesellschaft in Recklinghausen, the only topic of discussion at the Telegraphenberg was X-rays, the event was discussed in every room as if after an unexpected Nobel Prize announcement. He, however, did never publish a

---

[6] Kuiper (1946).

[7] Friedman et al. (1951).

numerical value for the high temperature of the corona that followed almost directly from his interpretation of the coronal red line.

At the beginning of August 1951, on the occasion of the World Festival of Youth and Students, his former boss Finley-Freundlich had travelled from Scotland to East-Berlin to ensure, with the help of the SED leadership, that Grotrian may again probe Einstein's prediction of the gravitation deflection of light during the solar eclipse in Sweden 3 years later. Freundlich wanted the observations "be repeated, preferably with the special telescope (horizontal camera) which was developed for this purpose and which has been in the custody of the Astrophysical Observatory in Potsdam since the last campaign in 1929." Academy director Naas wrote to Einstein on this question and received the cool reply that the previous determinations of light deflection have attained the degree of accuracy that is achievable with the present tools. "I doubt whether the progress due to a new expedition justifies the considerable costs of such a project."[8]

Grotrian who wanted to study only the magnetic sun and not the space around it, had also rejected the idea, saying that he did not feel at all qualified to work on the problem of light deflection. During the (unsuccessful) 1954 eclipse expedition to Sweden, Mattig and Strohbusch had concentrated on this subject. They had commissioned Carl Zeiss Jena to restore and improve the double-horizon camera for eclipse observations. Grotrian preferred to have a coronagraph built in his institute with which the magnetic eruptions of the sun could be seen. It was at this epoch that the academic sun was becoming increasingly magnetic, and he wanted to play a part in this concert.

A late summer party at the terrace of the Direktorhaus, almost everyone had come, scientists, assistants and the mechanics (Fig. 4.4), one of them barbecued for everyone. The front gardens of the residential buildings, which had been converted into vegetable gardens during the war, had been plundered. There was a babble of voices and laughter but suddenly Grotrian, in his patched grey-green knickerbockers and with his black cigarettes, asked: Where is Miss Starke? Nobody knew, someone whispered: Maybe in Werder? What's in Werder? Young people are being picked up there because of leaflets. "Miss Starke? She wants to study astronomy, that can't be right."

The next day, Grotrian had himself chauffeured to Werder. As always, he did not want to drive through the completely ruined city centre of Potsdam, but took the side route via Caputh near Einstein's summer house, which he had never seen from the inside (Fig. 4.5). They reached Werder with the village of Glindow on the left and the flashy villa on Chausseestraße. Alfred

---

[8] Hoffmann (2015).

Fig. 4.4 AOP 1951, farewell to H. Kienle to Heidelberg (1.10.1950); left to right: Czeschka, Borchert, Herzog, Brunnckow, Böcklein, Wempe, Starke, H. Strohbusch, Kienle, Grotrian, Daene, Künzel, Schneller, E. Strohbusch, incomplete. Kienle had carried out an enormous amount of reconstruction work in Potsdam in the post-war years but he hardly published science results any more. (Courtesy of H. Sperlich née Starke)

Fig. 4.5 Stadtschloss Potsdam 1946 where 1921 the first meeting of the Astronomische Gesellschaft after the WWI took place. Bombed 1945, dynamited 1959/1960 to stop the detailed discussions of its reconstruction, today fully reconstructed as site of the Brandenburg parliament (since 2015). (Photo B. Lüscher. Courtesy of Gisela Rüdiger)

Starke came down the stairs from the roof apartment and opened the door. That he knew nothing was immediately obvious: people would say that she had been arrested like dozens of young people in the neighbourhood. The father had searched everywhere in Potsdam and asked around, at the court and in the hospitals, even at the cemetery close to the observatory—and had received no information anywhere. Grotrian promised to find the daughter, to clear up the matter, to stand up for her, with the support of professors, directors, Academicians and national award winners; it could not be that in the state where he now lived, someone could simply disappear without leaving a trace.

Back in his office, a letter to the Mathematics and Natural Sciences Department of the Academy had been dictated: "I received the information that Helga Starke, employed at the Astrophysical Observatory, has not been here since Wednesday, 29 August 51 … Miss Starke is a member of the Society for German-Soviet Friendship. If there is any possibility to clarifying the case soon and possibly releasing Miss Starke I would very much welcome this in the interests of the continuation of the work at the observatory." After weeks somebody replied *"that we cannot do anything about this matter. Further employment is only possible once the case has been clarified. Of course, the salary payments must be stopped immediately."* Wempe had protected copies of these documents until his own end.

"Prisoners in GDR are always Guilty"

Shortly after the New Year's Day, Grotrian again met Mr. Starke but he still didn't know anything about his daughter's disappearance. He had not any information for months, not even at the headquarters of the Soviet Control Commission in Berlin-Karlshorst, where the father had dared to go. On 8 January, Grotrian wrote as the Director of the Observatory to President Friedrich[9]: *I informed the Academy on 3.9.51 that Helga Starke had been arrested on 29.8.51. The Academy replied that the Academy could do nothing in this matter and that salary payments were to be stopped immediately. Four months have now passed since this incident…In view of this situation, I feel obliged to draw attention to this case once again with reference to Articles 134 and 136 of the Constitution laws of the GDR. I am asking the Academy to take steps to clarify the case.*

---

[9] Walter Friedrich, President DAW 1951–1955.

Miss Böcklein had a lot of typing to do these days; the letter was also distributed to the Academy Director, the Secretary of the Mathematics and Natural sciences class, the advisor of this class and even the Personnel Section, Grotrian needed maximum publicity. At the end of January, he sent Prof. H.H. Franck all the papers in the case and wrote *I very much welcome the fact that you have also agreed to take an interest in this matter and I am grateful for the advice you have given. In the interests of my institute, the Academy and the opportunities discussed at the last faculty meeting to attract interesting colleagues from the West to our Humboldt University, I would greatly appreciate it if this matter could be dealt with and resolved correctly.* Franck, son of a well-known Berlin painter, early member of the SED and the Volkskammer from the very beginning, co-founder of the Chamber of Technology, Academy member since 1950, was the chair of chemistry at Humboldt University and director of an Academy institute. Grotrian sought support wherever he could find it. At this time Helga Starke was already on the long way to one of the camps in Workuta north of the Arctic Circle in the middle of winter.

On the same day in January when Grotrian dictated his letter, the first secret Werder tribunal took place in the building at Lindenstraße 55. In the oak-panelled lower hall, a table covered in green cloth separated the prosecutors from the four female defendants, the latter unwashed in rough remnants of uniform, guarded by heavily armed soldiers. The women, among them Helga Starke, still hoped for the end of their suffering as nothing had happened that they could have been accused of. There was an interpreter and only the "Russian troika": a judge with two assessors. There was no defence counsel. The accusations varied slightly, in Starke's case it was propaganda, agitation and dissemination of writings, one more charge than the neighbouring medical assistant Geske. She was accused of transporting two letters, Helga six. The court took such differences indeed into account: Helga received 25 years in a Gulag camp, Geske 15 years. At the next trial, on 11 January, the defendants (three men and a woman) were given death sentences, all of which were executed in Moscow.[10] Such sentences would have been completely illegal, the Russian side will state in the rehabilitation letters after 1990, because the defendants had not harmed the USSR at all and all the trials had not taken place on Soviet territory.

In April 1952, Director Naas reported to Grotrian that he had a telephone conversation with the Minister for State Security, Mr. Zaisser, who *answered all inquiries satisfactorily.* The Minister *herewith informs Prof. Grotrian that the assistant had been arrested in order to carry out an investigation and that,*

---

[10] Spiegel (2002).

according to the status of the investigation, Prof. Grotrian could not count on the accused person for length of time. In response to the reference to the Constitution law of the German Democratic Republic the Minister also added the general remark, namely that no innocent person has ever been arrested in the German Democratic Republic, nor has anyone who has been arrested been detained after being found innocent. Naas added that he had thanked the Minister for his personal words and that the information he had received was completely sufficient. This should teach the prominent protester that there was nothing more to ask, in particular not in the future. Four weeks after these final formulations, Mrs. Starke arrived in one of the northern work camps after weeks of transport, someone had slipped her a coat on the way. The female prisoners of Workuta had to ballast freshly laid railway tracks; the brigadier, a young German-speaking Russian woman, remained her only contact person during the endless forlorn year 1953.

From Munich, from the 1952 meeting of the Astronomische Gesellschaft, Grotrian telegraphed to the Academy President on September 23: *May I invite the Astronomische Gesellschaft to the 1953 autumn conference in Berlin?* On the same day he received that *I agree to the 1953 autumn meeting of the Astronomische Gesellschaft in Berlin. Conference venue: Academy of Sciences*, the AdW still saw itself as an all-German research community. The next day, Grotrian offered the now official invitation from the Academy to hold its 1953 meeting in East-Berlin. "The invitation was received with great approval"[11] and he was re-elected to the board of the Astronomische Gesellschaft.

During the preparatory discussions with the Chairman of the society, Otto Heckmann in Hamburg, the real problems had been formulated. Heckmann demanded that "the modalities … be organised in such a way that the members decide to participate in sufficient quantity." Still in March, Grotrian informed the presidium of the consequences of this condition for the realisation of the conference. Firstly, the travel procedures should not be too inconvenient, which would require special instructions to the border police, and secondly, the costs should not exceed the usual costs for participants at meetings in western cities, as otherwise too few members would come. The latter condition could hardly be met without additional payments, not least because the planned excursion to Potsdam required additional expenses that were not easy to predict. The exchange modalities of the two currencies also resulted in a complicated calculation for the total requirement of around 8000 marks (east) including pocket money, which would have to be provided by the Academy. Grotrian informed *that the president of the A.G. would only ask its*

---

[11] Mitteilungen der Astronomischen Gesellschaft 4, 5 (1952).

members to hold the conference in East-Berlin if these demands were fulfilled. The presidium indeed agreed to bear all the costs and to send the necessary residence permits by mail in good time: best prospects for a successful meeting of all German-speaking astronomers in Potsdam by an afternoon tour from Sternwarte Babelsberg via Sanssouci to Telegraphenberg. Grotrian had dozens of questions to answer in Potsdam: Scheduling, evening lecture, slide projector, microphones, restaurants with the different price categories, where will the VIPs stay, who organises the ladies' programme, how to report on the considerable reparation losses of instruments and books of both institutes?

On May 1953, in addition to the ongoing preparations for the solar eclipse expedition, the observations started with the guide tube of the new coronagraph. At the same time, the organisational preparations for the autumn conference of the AG were in full swing. At the beginning of June, the concert programme designed by the orchestra musicians of the Komische Oper was printed, but suddenly political turbulence captured East-Berlin and the GDR. After Stalin's death in March, his successors had dissolved the Soviet Control Commission and ordered Walter Ulbricht to Moscow to stop his separatist course towards Germany, his absurd church struggle and the system of economic impertinences. The people found out about an alleged New Line and the associated self-criticism of the party leaders—without resignations—from the newspapers on the morning of 12 June; some orders were indeed withdrawn, but not the hated standard norms for construction workers.

Until 17 June, there had been almost daily riot and strikes, initially in the countryside or in small towns, also near Potsdam. By the 17th, more than 10,000 people were protesting against the government in the centre of Brandenburg/Havel with demands for its resignation. The buildings of the state party SED and the district court were besieged, the prison and the police office were partially occupied. Shots fired at the demonstrators by the Soviet Army put an end to the revolt. In East-Berlin and in large cities such as Halle, Leipzig and Dresden, tanks were deployed to cordon off streets and squares. The political demands required by the demonstrations were for free elections, the release of political prisoners and the reunification of the two Germanies. The international press had carried the images of the tanks in East-Berlin all over the world, the city resembled an army camp in the eyes of the reporters. Around 20,000 soviet soldiers, 800 tanks and 15,000 barracked police occupied East-Berlin. At 2 p.m. on 17 June, the Soviet High Commissioner radioed to Moscow that the demonstrators dispersed with the appearance of Soviet tanks (Fig. 4.6).

On Friday, 26 June 1953, Grotrian made a call to the President in the afternoon. He had unexpectedly *received a telegram from the chairman of the*

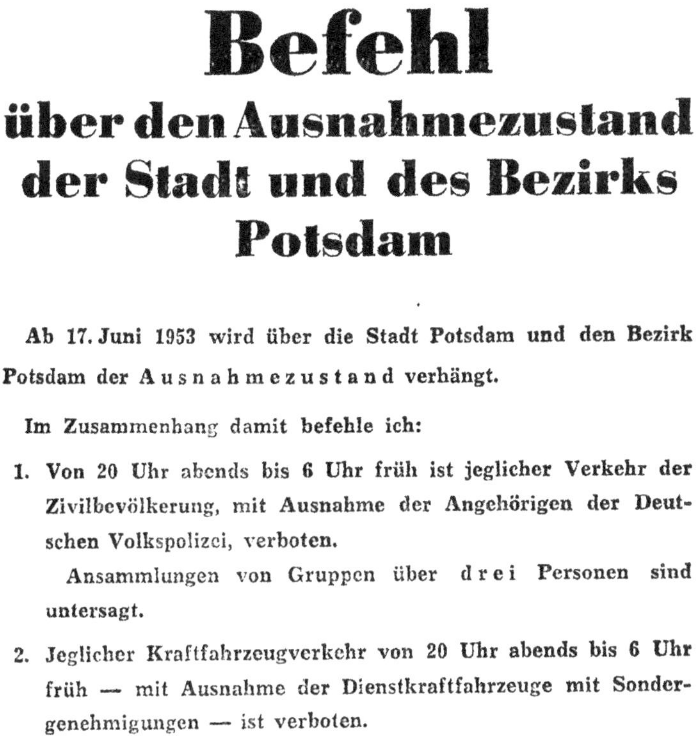

Befehl
über den Ausnahmezustand der Stadt und des Bezirks Potsdam

Ab 17. Juni 1953 wird über die Stadt Potsdam und den Bezirk Potsdam der A u s n a h m e z u s t a n d verhängt.

Im Zusammenhang damit befehle ich:

1. Von 20 Uhr abends bis 6 Uhr früh ist jeglicher Verkehr der Zivilbevölkerung, mit Ausnahme der Angehörigen der Deutschen Volkspolizei, verboten.

   Ansammlungen von Gruppen über d r e i Personen sind untersagt.

2. Jeglicher Kraftfahrzeugverkehr von 20 Uhr abends bis 6 Uhr früh — mit Ausnahme der Dienstkraftfahrzeuge mit Sondergenehmigungen — ist verboten.

Fig. 4.6 State of emergency on 17 June 1953 proclaimed by the Soviet commander for Potsdam: "(1) from 8 p.m. to 6 a.m. all civilian traffic is prohibited. Gatherings of groups of more than three people are prohibited. (2) All motorised traffic from 8 p.m. to 6 a.m. is prohibited." In the Potsdam district, 575 people were arrested; the status lasted until 29 June, in East-Berlin until 11 July. (Courtesy of Potsdam Museum)

Astronomische Gesellschaft in Hamburg stating that the meeting of the Astronomische Gesellschaft in the Democratic Sector of Berlin was to be postponed to another year. The file noted by quoting him that this was the end of the matter for him. The meeting that year took place in Bremen at the beginning of October. The Academy management yet still approved currency travel funds for nine east German astronomers. The annual report of the AOP for 1953 states that the meeting of the Astronomische Gesellschaft, which was planned to take place in Berlin and Potsdam and whose preparation had required a considerable amount of work, "was attended in Bremen by 12 members of the observatory with the exception of the director, who was sick."

After Stalin died in March 1953—mourned in East Germany with millions of black textiles hung outside windows—conditions in the Russian prisoner camps improved. H. Starke was even released completely at the beginning

Fig. 4.7 Solar eclipse expedition Öland (Sweden) 1954; left to right: Schürer, Freundlich, Mattig, Wempe. (Legacy Mattig. Courtesy of W. Schmidt)

of 1954 and transported back to Potsdam, but after a short visit to the observatory she decided to go to West Germany which she was indeed free to do. Walter Grotrian had fallen seriously ill during the AG meeting and had to go to St. Josefs Hospital in Potsdam[12]; he died there early in 1954, before his institute's great but unsuccessful solar eclipse expedition to Öland (Fig. 4.7).[13] The funeral procession at the municipal cemetery on the Telegraphenberg was unmanageably long, writes Wempe in his obituary.[14] Grotrian's master students Mattig and Schröter published the last magnetic field results posthumously under his name[15] and left Potsdam and the GDR just in time at the beginning of August 1961. "On Saturday evening we celebrated the successful crossing to the West at Ingrid's brother's house. I don't want to use the word escape for this action, because the Stasi wasn't right on our doorstep. The party went on until the early hours of the morning, one of the chairs broke, and the next morning we were woken up with the news that the border between West- and East-Berlin had been completely closed. That was 13 August 1961," reported Mattig in his memoirs.[16] The friends will not see the location of their first years again until 29 years later, when they meet the current solar researchers from Telegraphenberg at a conference of the

---

[12] Kienle (1956).

[13] The event could be observed quite well in Potsdam as a deep partial eclipse (88%) under clear sky.

[14] Wempe (1955).

[15] Grotrian (1956).

[16] Dick & Ackermann (2025), p 283.

Astronomische Gesellschaft in Berlin in spring 1990 and spontaneously travel to Potsdam. Jürgen Rendtel will guide them through the Einstein Tower and Mattig will realise that very little has changed.[17] At the opening ceremony in 1992 of the newly founded Astrophysical Institute Potsdam, he will give the ceremonial lecture in the dome of the Great Refractor.

Stillstand

After Grotrian's death, the observatory was managed by a short-lived Scientific Board headed by Wempe as the chairman. The independent Academy President Friedrich, radiation physicist and Sommerfeld student, had high expectations of the scientific excellence of the directors of his research institutes and tried to avoid inhouse appointments. As Grotrian's successor he could only imagine an astronomer with an international reputation. He headed a search commission of varying composition and with a well-equipped list of candidates, probably written by Kienle.[18] The West German names, however, disappeard from the final list, after the East German slogan "Germans at one table" had become obsolete. It originally contained names from Germany, Western Europe and USA. Copies of the presidential-letter to Brück (Dublin) from May 1954 as well as the reply letters from Brück,[19] Chalonge, Minnaert, Schwarzschild, Swings and Wildt have been preserved. In addition to the cover letter, Appendix 5 contains the replies from Brück, Minnaert and Schwarzschild as well as a corresponding letter from Klüber dated January 1956 to Kienle who had apparently approached him to bring him back to Potsdam after receiving all the negative replies. All the candidates had finally cancelled, Martin Schwarzschild with the explanation that it would be a great honour to be asked to direct his father's former institute, but who would ever want to leave Princeton if he had already settled in there? Certainly not, as the rejections can be read throughout, after Soviet military tanks had recently rolled into East-Berlin ensuring the GDR's celebration of its fourth birthday.

I enclose a copy of a letter that I received recently from Mr v. Klüber in response to my enquiry, Kienle wrote to the chairman of the search committee, on 4 January, and *If things in the GDR continue to develop as indicated by the latest*

---

[17] Personal communication Rendtel.

[18] Brück, ten Bruggencate, Chalonge, Haffner, Minnaert, Schwarzschild, Siedentopf, Swings, Wellmann, Wildt.

[19] Hermann Brück (1905–2000), Daniel Chalonge (1895–1977), Marcel Minnaert (1893–1970), Martin Schwarzschild (1912–1997), Pol Swings (1906–1983), Rupert Wildt (1905–1976).

show trials, then it will probably not be possible in the long term for anyone who loves justice and freedom to work in and for this state. This was unambiguously so that the finding phase ended immediately and without result, also the time of President Friedrich had expired. *It is not possible to recruit a researcher from the Western democracy to head an institute of the DAW in Berlin,* Kienle wrote to Friedrich in February 1956 on the situation of GDR astronomy in his Memorandum on the situation of astronomy in the GDR.[20] The call for applications was cancelled and Grotrian's absence representative Wempe became appointed director on 14 June 1956.

Only one of the candidates, Hermann Brück had wanted to see the situation in Potsdam, the site of his first post-doc years on the Telegraphenberg. Now he gave a seminar lecture and took part in a meeting on the 2-m-telescope—whose future location was still open—to which Kienle had also travelled to. The Potsdam of 1955 lay in ruins, the ruin of the Garnisonkirche—in front of which thousands of Potsdamer's had practised welcoming their new leaders in March 1933, first a few, then many, finally a sea of upraised right-wing arms—the remains of the destroyed Stadtschloss. The local astronomers remained silent, they knew who is visiting, what this finally could imply. Brück was one of the preferred students of Sommerfeld with the opportunity to know him on the ski slopes. "He was an expert skier and had a weekend house in the mountains where his assistants and senior students, as well as occasional foreign visitors—ten or so people altogether— would be invited. In those days we carried our skis on our shoulders, and trudged all the way up the icy track."[21]

Since December 1928, Sommerfeld and Freundlich had organized a grant from the Notgemeinschaft der deutschen Wissenschaft and—after the external funding expired—Brück was employed by Ludendorff on 1 September 1933 for life, backdating his qualifying age to 1930. A productive, professionally careless time, he writes in his biography, they played tennis in the grounds, spent the night somewhere in the Great Refractor building over the week and had lunch together in the lecture room, delivered from a Potsdam restaurant. The director Ludendorff rallied his followers around him against Freundlich. Brück was catholic but his girlfriend was Jewish, ruling out a common future in Germany after the Nuremberg Race Laws of September 1935. It became more and more clear that he would have to choose between family and career in Germany. It was not without ulterior motives that Brück visited the Vatican Observatory in Castel Gandolfo for 4 weeks in the autumn of 1935; when

---

[20] Dick & Ackermann (2025), p. 247.
[21] Brück (2000).

astronomers have to leave, they will always try to reach another observatory. Brück escaped his home town Berlin with Irma Waitzfelder on the last days in July 1936 shortly before the opening day of the Olympic Games (Fig. 4.8).

Ludendorff wrote in the 1936 annual report that the scientific assistant Dr. Brück left his position here on 1 August. Because of more lax border controls to Austria because of the Olympic games, both refugees reached Rome undisturbed, where they married in September and the young husband worked at the Vatican Observatory for some time, almost unpaid, before being employed by Eddington in Cambridge the following year. In late summer, Ludendorff

Fig. 4.8   Hermann Brück (1905–2000) and Irma née Waitzfelder (1905–1950) with their children Maria and Peter. (Photo A. Beer. Courtesy of P.M. Bruck)

wanted to lure him back to Potsdam *for the purpose of correct dismissal*, but Brück had not been so naive as to fall for it. He wrote to the *esteemed professor* that he would be *particularly grateful to him if I could settle the termination formalities by letters*. He will write to Rolf Müller and ask him to take his few personal matters into his own hands. Ludendorff replied that *your abrupt departure from your position is a completely incomprehensible behaviour for which I have nothing but the strongest disapproval. Heil Hitler.* But with *Since Dr. Brück has resigned … he is to be considered dismissed from the work as of 31 July 1936* the Ministry of Science finished the dispute over the *scientifically highly talented man … who had enjoyed a high reputation with me and all his colleagues* (Ludendorff).[22] In Potsdam Brück had conducted spectral statistics with 17,000 stars in the southern hemisphere with data from the Potsdam Spectral Survey.[23] He wanted to know whether the sun is located in a star cluster or even a stellar hole and found that the B stars and, to a lesser extent, the A-type stars show a stronger galactic concentration and "that the sun is probably not located in a very pronounced star cluster." Data on interstellar absorption were not yet available, and their consideration led to a partial revision of his views a short time later. Nevertheless, it was evident that the density is lower everywhere up to distances of about 2500 pc than in the neighbourhood of the sun. The sun must therefore be located in a star-rich region of the Milky Way system, Wilhelm Becker later confirmed the former colleague.[24] In his detailed obituary for Ludendorff Brück formulated that "it was not always possible to agree with Ludendorff's views, but his generous and straightforward character made him greatly respected everywhere."[25]

Wempe was a member of only a very few of the countless political Nazi organisations, i.e. the so-called Nationalsozialistischer Lehrerbund NSLB and the Nationalsozialistische Volkswohlfahrt NSV starting 1935/1936. He was appointed professor at the Brandenburg State Highschool since 1948 with a full teaching position and a considerable honorarium. In a personal curriculum vitae dated 7 August 1948 he wrote that as a result of difficulties "caused by my non-membership of the NSDAP, I gave up my regular position in Heidelberg in 1938 and took on a nonregular assistant position at the Universitätssternwarte Jena. However, I had to refrain from applying for a lectureship as I was not a member of any Nazi organisation." He therefore left

---

[22] ABBAW, AOP, No. 149.

[23] 60,000 stars up to 13th magnitude in the Kapteyn Selected Areas (Becker, 1929). The photographs were taken in 1927/1928 in La Paz (Bolivia) with a Zeiss triplet of 30 cm aperture (spectrum and brightness). Observers: Becker, Kohlschütter, Müller.

[24] Becker (1942), Friedrich and Wilhelm Becker were brothers.

[25] Brück (1942a, b).

the university and moved to the Potsdam Astrophysical Observatory where he was able to work without teaching duties. In Potsdam he had, in order to continue the institute in Grotrian's sense, later transferred the concept from the magnetic sun to the magnetic stars and appointed both the assistants Mattig and Schröter to senior assistants. However, the 2-m-telescope, which at Kienle's request of 11 April 1949 was already in production in Jena, had been taken away from the new director Wempe and the AOP (Fig. 4.9). It had suddenly be transplanted to Tautenburg/Thuringia for a newly founded independent institute. For the candidates succeeding Grotrian the Academy President had still promised on 31 May 1954 that a 2-m-mirror telescope is planned for the Astrophysical Observatory. However, the planned location in the Fläming mountains south of Potsdam was not approved, Wempe wrote suggestively without giving more details in the annual report for 1956. In his application for the telescope, Kienle had avoided all location details, but favoured the foundation of a Central Institute for Cosmic Physics within the

Fig. 4.9 2-m-telescope, direct focus 4 m, focal length in Cassegrain focus 27 m, focal length in Coudé focus 92 m. Schmidt system with wide field of view. Two guide tubes built into the main tube (Kienle, 1949)

Fig. 4.10 Astronomers in the mountains near Potsdam looking for a location for the 2-m-mirror telescope; left to right: unknown, Künzel, Strohbusch, Kahrstedt, Grotrian, Daene. (Courtesy of Förderverein Großer Refraktor Potsdam)

DAW, consisting of the "remnants" of the AOP, the Babelsberg and Sonneberg observatories and the Astronomisches Recheninstitut.[26] The Ravensberg near Potsdam was the favoured place after Kienle's departure (Fig. 4.10). In the Memorandum on the situation of astronomy in the GDR of 1956, Kienle had warned that *the 2-m-reflector telescope must not [be taken up] under any circumstances… [as] a replacement for the dismantled Babelsberg reflector …. because at present we do not have a single instrument that can be used for spectrographic work.*

In 1957, the AOP received a new 70-cm-mirror telescope including an observation house for light-electric observations of variable stars by H. Schneller and a facility for a spectroheliograph to the south of the refractor building. Neither installation had a long life because of a missing clever working plan in these years. In his letter to Kienle, Klüber had warned that *today in Potsdam a whole group of ambitious young people are apparently muddling around without proper leadership.*

---

[26] Kienle (1949).

Fig. 4.11 Left to right: Friedrich W. Jäger (1914–2000) at the top position of the AOP (ca. 1960); Heribert Schneller (1901–1967). (Courtesy of B. Eggers/Förderverein Großer Refraktor Potsdam)

On 1 April 1957, Wempe hired the polarisation expert F.W. Jäger who had completed his doctorate with ten Bruggencate in Göttingen, on a private contract as head of the Einstein Tower and Professor of Solar Physics with a full professorship at Humboldt University (Fig. 4.11). "He supported us in carrying out the routine work," Mattig writes coolly in his autobiographical report; there were no joint scientific projects of the new professor and his senior assistants. The solar tower, however, was soon assigned to the newly founded Central Institute for Solar-Terrestrial Physics (ZISTP), which challenged the astonished Jäger with unexpected requests including questions such as whether, in addition to the effects of solar activity, there are also effects on the living conditions on Earth or—more simple—how much solar energy differently oriented photovoltaic systems can be collected from the sky.[27]

Wempe liked to make work easier for people. He created additional working space in 1961 by glazing the two arcades of the main building. Scruples about this intervention in the traditional Telegraphenberg architecture never completely left him; today's ambitious conservationists would probably scold him, the original architects probably not.

---

[27] Jäger (1972, 1979).

5

Astronomy in Borderlands

Sonneberg in the Frontier Area

On October 1962, Cuno Hoffmeister, director of the Sternwarte Sonneberg, complained to his local district council about the disadvantages that the observatory suffered from its location in the 5 km restricted area (Fig. 5.1). "Since October 1, 1961, the Observatory of the Academy of Sciences, which

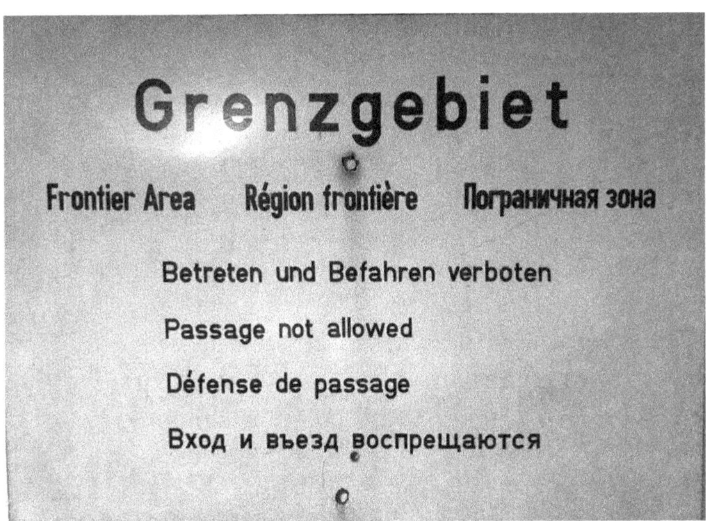

Fig. 5.1 Standard warning for frontiers areas along the GDR border fortifications: no trespassing or travelling

G. Rüdiger, *The Astrophysical Observatory Potsdam - Triumph and Tragedies*, Astronomers' Universe, https://doi.org/10.1007/978-3-032-06294-9_5

now had the largest collection of photographic images of the sky in Europe, had been located in the restricted border zone, only 300 m from its outer edge." Hoffmeister avoided the correct term "state border"—which the readers of his letter will have noticed first of all. He reported that he had been trying to interest the leaders of the Academy of Sciences (DAW) since last spring and had received an urgent recommendation from them to contact the local authorities about the details. He now asked for at least the admission of around five visitors a year also from the Federal Republic of Germany and Western countries on the condition that they do not leave the institute ground for the duration of their stay, so that they have been accommodated and fed. Hoffmeister wanted to do all this himself, certainly with his own money since no price was too high for him to contact astronomers personally from all over the world. In July 1963, Hoffmeister rejected a form from the Academy for the annual registration of conferences: *Due to the location of the observatory in the 5 km Restricted Zone, it is currently senseless to plan conferences.* At the beginning of 1963, Hoffmeister reports the visit of his colleague Lukas Plaut from Groningen for one day to the Academy headquarters: This is the first time we have succeeded in breaking through the strict border zone regulations. It was not until 1972 that the Sonneberg city area—and only this area—including the Erbisbühl hill, on which the observatory is located, was released from the zone, too late for Hoffmeister and probably also too late for his observatory.

The Sternwarte had experienced an enormous progress after 1955 and the number of buildings, instruments and employees had risen continuously (Fig. 5.2). The 40-cm-astrograph, which had been dismantled in 1945 as a result of the war, had been replaced twice in good quality; Hoffmeister financed the somewhat smaller one with his own money. Prior to this, Wenzel, Richter and Jackisch each spent 6 months at the Heidelberg Observatory working on the Sonneberg Field Plan with the 40-cm-astrograph there, providing new food for the discovery hunger of their boss, who had found by himself more than a third of all variable stars known to astronomers at the time. Construction of the large new main building with astrodomes 5 and 8 m in diameter and a metal hut on the roof for monitoring the sky had already begun in 1957. The observatory was now not only the highest observatory in Germany, but also one of the best equipped.

Hoffmeister's position was still untouched in May 1961. On May 12, he received the information that the Academy in Berlin would cover the costs of the business trip for him and Nikolaus Richter to the 11th General Assembly of the IAU in Berkeley, USA in August and—at Hoffmeister's request—would provide travel authorization for his colleague Schubart, albeit without a

Fig. 5.2 Sternwarte Sonneberg, the staff 1963; left to right: Wenzel with son Hans, Thänert, Bartl, Kraus, Wicklein, Heim, Hoffmeister, Lehnhausen, Götz, Geißensetter, Häusele, Eichhorn, Brandt, Huth, Heimann, Ahnert, Rößiger, Ziegler, Kalbe, Bräuer, Richter. Missing: Fürtig, Jackisch, L. Meinunger, Rose. (Photo I. Häusele. Courtesy of W. Fürtig)

financing commitment. On May 18, a letter was sent from the Academy President to Sonneberg stating that it had been decided to elect the corresponding member Cuno Hoffmeister, Director of the Sonneberg Observatory, as a full member. Kienle handwrote the following rationale: I can only add to the laudation I wrote a few years ago that Hoffmeister, as he approaches 70, is increasingly impressing us younger people with his unlimited creative power, which allows him to use the nights for observation, the days for evaluation and any free time in between for talks. Hoffmeister was the only GDR astronomer (except Kienle) to remain as an Academy member, appointed before the time, when non-members of the SED were negligible as candidates. On the occasion of his 70th birthday on February 2, 1962, J. Hoppe described the life of the jubilarian: In the solar system, Hoffmeister began with the study of meteors (1910) and zodiacal light (1912). The space of our stellar system was entered with the variable stars (1913). Finally, in connection with the last African expedition (1959), the advance into extragalactic space was made through the discovery of a hitherto unknown object, an extragalactic dark cloud.[1] Indeed, at the joint meeting of the extragalactic astronomers and radio astronomers in Berkeley, Hoffmeister emphasized his exceptional position in German astronomy with a contribution on this extragalactic dark cloud in the star constellation Microscope which would have the dimensions of the Magellan Clouds. In this area the number of brighter galaxies was greatly reduced, while that of the RR Lyrae stars would have hardly changed.[2]

The IAU General Assembly 11 was characterised by a new cooperation between East and West. In his closing speech, the Soviet astronomer Ambarzumjan—who had just been elected as the new chairman of the IAU—required the strengthening of scientific exchange between East and West: "I would like to hope that all forms of scientific communication, including the exchange of scientists, will be expanded." Hoffmeister wrote in his travel report that in addition to 319 US astronomers, 64 Soviet and 50 German astronomers had taken part, whereby he had simply added those from the two German states together. The beauty of the country, the balanced climate, the exemplary organisation of the events and the collegiality of international cooperation, emphasised at a time of great political tension, created an atmosphere of peace and community that was felt joyfully and gratefully by all participants, he reported the meeting and his own political programme. Science also benefited greatly from the "Tauwetter" after Stalin's death. In the GDR, this period came to an early end already in 1964 when the director of the Institute of Physical Chemistry, Havemann, lost all his titles, posts and employment within a short time for his famous lectures about Scientific Aspects of Philosophical Problems.

The Californian astronomical world festival was interrupted, possibly with some delay, by the political news of the construction of the Berlin wall, first with barbed wire, then with stones and later with impassable concrete. The Ulbricht regime had cancelled Berlin's four-sector status overnight and declared East-Berlin the capital of the GDR. The very last life insurance of East Germans, getting on the city train in Potsdam and getting off as West German citizens after three stops in Wannsee, was over, the bottleneck closed. Although only a handful of the participants in Berkeley will have been directly affected, the question of the political reactions will have interested many, especially because of the military tanks at Checkpoint Charlie and the lack of reaction from the young President Kennedy, who had just taken office and only after a few days had sent his Vice President with General Clay to Germany. Almost unnoticed during these days was the military fortification

---

[1] Hoppe (1962).
[2] Hoffmeister (1962).

of the inner-German border, combined with new regulations on the use of fire weapons as "escape prevention." Any border crossing became dangerous for life. A 5 km wide Restricted Area—with a 500 m wide security strip close to the border and a 10 m wide death strip—had been established, which could only be entered with special authorisation, the issuing of which took an unpredictably long time due to the large number of official security officers. Unfortunately, the Sternwarte Sonneberg was located just inside the restricted zone, so visitors from West Germany had to fight with many special difficulties.

The celestial mechanic Schubart, Hoffmeister's man for orbital determinations of all kinds, had received travel support from the AG and the IAU to attend the General Assembly and had also exchanged private money for convertible DM at the usual rate of 5:1, with which he was able to pay for the flights to San Francisco but not the journey home. In 1955, he had successfully applied in Sonneberg for the open position of the deceased meteor researcher Eva Ahnert-Rohlfs and had proved himself in Hoffmeister's eyes; the unpolitical Schubart was soon promoted as his "presumptive successor" in Sonneberg, albeit unconfirmed by state authorities as Schubart was not even a member of the official GDR labour organisation. In January 1959, it was decided by the Academy authorities that Hoffmeister would "not be legally retired until further notice and that he would propose a suitable successor in the foreseeable future." The hard-working Schubart had quickly determined the orbital elements of comet C/1959 01 discovered in 1959 on Hoffmeister's plates taken in South Africa.[3] In Berkeley, Schubart attended sessions on celestial mechanics and the solar system. He took part on the extensive programme of observatory visits at Lick, Mt. Wilson and Palomar, but the return journey made everything even more impressive. Schubart had arranged a free passage on a cargo ship from Houston to Bremerhaven, with wheat taken on board in New Orleans. This first required a long bus journey along the Californian coast to the Mexican border and through the deserts of Arizona to Tucson with the Steward Observatory of the local university and the city station of the new Kitt Peak National Observatory. Then he travelled by plane to Texas, across vast landscapes under a bright sky. In contrast, the tiny Sonneberg, located in the dark Röthen valley, was closed in all directions: northwards by the Thuringian Forest and southwards by the well-guarded border fortifications, the sky was often overcast, even the high observatory only managed a little over 100 observation nights a year. The return to Hoffmeister's observatory would not be tempting after 13 August, Schubart

---

[3] Schubart (1961).

must have thought on the waters of the Atlantic Ocean. The West German ship's officers provided their exotic East German passenger with the latest radio news on the situation in and around Berlin. Schubart guessed that the Astronomische Recheninstitut in Heidelberg was looking for a celestial mechanic; he decided not to return to Sonneberg after his arrival in Bremerhaven. Cuno Hoffmeister received the final letter on 4 October 1961: "I am very sorry to have to disappoint you: I have decided to stay in the Federal Republic of Germany. I am therefore not returning to Sonneberg and am hereby resigning from my position at your institute." This letter came from Hamburg, and his later conference report indeed noted Astronomisches Recheninstitut Heidelberg as the author's address.[4] Hoffmeister in his annual report for 1961 wrote that the institute had suffered a very serious loss, all the more so as Schubart had started to habilitate after consultation with Lambrecht at the Jena University Observatory in order to pursue the academic connection of Sonneberg astronomy alongside Hoffmeister, which later ceased to exist completely without the two protagonists.

Reunion in Bamberg

In Bamberg, Hoffmeister—as a young amateur—took up the temporary position of an astronomer at the observatory who had been called up for military service in April 1915. On behalf of Director Hartwig, he observed meteors and variable stars and learnt early on what it meant to be an astronomer. As a young man, he sometimes worked on the instruments for 12 h on cold winter nights … and even as a 70-year-old he hardly missed a clear night, Hoffmeister later wrote as a student in the obituary of his teacher, which sounds as if it were his own. "The most beautiful thing about him, however, was his pure and great enthusiasm for his profession, which already inspired him during his time in Strasbourg and did not leave him even in the last year of his life."[5] The young Hoffmeister had transferred Hartwig's interests in interplanetary space and the variable stars from Bamberg to his own observatory on the 638 m-high Erbisbühl.

"Almost at the last minute, I would like to ask the presidium [of the DAW] to make it possible for members of the Sonneberg Observatory to attend the international colloquium on variable stars in Bamberg from 4 to 7 September 1962. The three lecturers Ahnert, Hoffmeister and Wenzel are the first to be

---

[4] Schubart (1962), Dick & Ackermann (2025), p. 247.
[5] Hoffmeister (1923).

considered, but equally important are those senior members of staff who are actively working in the field," Hoffmeister wrote to Prof. Dr. Klare on 20 August 1962. Two weeks before the beginning of the conference no prospect of the travel visas had been received in Sonneberg. The GDR bureaucracy denied him this journey to his roots, to a presentation of his achievements in his original work place, without giving any reason. In his crystal-clear handwriting, he had added to the request to be allowed to travel to Bamberg: "I would like to point out once again that Bamberg offers a first-class international programme. It would be a real detriment to our work if we were unable to attend." A last word from the petitioner, a recognisable gesture of subjection, without success. There were reasons for such decisions, he was told without further explanations. The great Hoffmeister, director with an individual contract and official head of all astronomers in the Academy, was banned from travelling and had been chained up like the other astronomers.

In 1959 Hoffmeister worked at the Boyden Observatory in South Africa. There he recorded thousands of visual measurements of RR Lyrae and RW Aurigae[6] stars and exposed almost 1000 photographic plates. For him, RR Lyrae and RW Aurigae were the protagonists of the two large families of variable stars—known today as pulsating and eruptive variables respectively. Hoffmeister's planned contribution to the Bamberg Conference in 1962 was to consist of a discussion of the discovery probabilities of the RR Lyrae stars on the photographic plates that he had taken to reveal the structure of the galactic halo in South Africa. He defined theoretical discovery probabilities and tested the values with his own data. With all his experience, he had found just half of the probable stars. Nevertheless, this seemed sufficient to establish that neither the distribution nor the mean brightness of the RR Lyrae stars had any relation to the dark cloud he had discovered in the star constellation Microscope. His paper concluded that the RR Lyrae stars of the galactic halo clearly belong to the foreground of the dark cloud. He had sent it by mail to Bamberg, where it was presented by a participant from Hamburg. The next year, in his public annual report for 1962 he had written that Wenzel and he had taken part in the Bamberg conference, albeit in absentia. Schubart from Heidelberg, who was present in Bamberg, will have seen no reason to doubt his previous year's decision.

The next Bamberg conference on "Position of variable stars in the Hertzsprung-Russell diagram" took place in August 1965. In the meantime, astrophysics with its theory of star formation had taken over the variable star research, which had led to a very high number of participants. 90 astronomers

---

[6]Today: T Tauri stars (Hoffmeister, 1957).

from East and West discussed the physical origin of known variable stars, the phenomenon of the U Gem stars, the symbiotic stars, all topics that will come to dominate stellar physics over the course of the next decades. Hoffmeister's former assistant Kippenhahn started with the key note lecture. The latest stellar model calculations regularly yielded overstable pulsating stars when a fixed temperature of 7500 K is reached in the giant stage. The oscillation frequencies and the other stellar parameters corresponded to the observed values for δ Cepheids.[7] The stellar evolution code did not yet go far enough to treat even the short-period RR Lyrae stars in a similar way, but Baker in Munich had fed his computer program with the known parameters of these old stars and obtained the observed run of the pulsation frequencies with the star colour in globular clusters.[8] In August 1965, Bamberg was one of the most beautiful places in the world for an astronomer.[9]

Hoffmeister had fought—only 4 years after the Berlin Wall—in numerous applications and petitions for permission of 11 GDR participants to attend the conference, including 9 from Sonneberg. In May 1965, he protested to the Department of foreign affairs of the Academy that he should travel alone to Bamberg for the IAU colloquium on variable stars. *I would also like to point out that the financing … will be provided by the organiser. I therefore see no reason to deny participation to someone who devotes most of his energies to the field of work being discussed,* he argued. Not a single other director of a GDR academic institute, whose official trip to a western country was about to be approved, would have even dreamed of campaigning for his colleagues. When he wrote *that only that group of scientists can hope to maintain its standard which works in direct contact with the world-wide progress,* he was also relying on the power of the argument. Such letters could still be written without political punishment but less than 2 years later the wind would change.

This time Hoffmeister's persistence had led to success. A large delegation with 7 astronomers from the Sonneberg observatory were able to travel to Bamberg. Hoffmeister's victory over the scientific bureaucracy had signalling effects, because if only 4 years after the Berlin Wall, many East German astronomers could meet with their western colleagues, then the consequences of the Wall on science could possibly remain manageable. Also, the (last) common AG conference in Eisenach in September 1965 seemed to

---

[7] Kippenhahn (1965).

[8] Baker (1965).

[9] Hoffmeister (1965).

reinforce this misinterpretation. Never again until the end of the GDR would a Sonneberg astronomer be able to reach a conference or an institute in the west. The observatory would remain cut off from all external developments. The directors in Potsdam were only interested in tetrad theory, non-Lorentzian transformations and maximally the rotation of the cosmos, but they were not interested in the light curves of real stars. However, none of these leaders ever achieved an international resonance with any of their publications as, for example, Ludwig Meinunger with his final determination of the rotation period of the real symbiotic binary star system AG Draconis.[10]

Hoffmeister had demanded that people from Sonneberg present their own papers in Bamberg. Towards the end of the proceedings there are short reports by Ahnert, Thänert and Wenzel. His own presentation was peculiar. Hidden behind the tricky question of how to deal statistically with years of data loss, he talked in detail about the loss of his 40-cm-astrograph in 1945 "which was called demontage" which he had been working with since 1938. He had to wait until 1961 for a replacement instrument of the same quality, so that the old plates could actually be compared with the new ones (Fig. 5.3). Did he want this dramatic break in his astronomical life to become known all over the world? Ahnert and he were once members of the NSDAP[11] and were not entirely uninvolved in the events that led to the reparations which further demolished East-Germany in almost all areas.

Fig. 5.3 The 40-cm Zeiss astrograph from the early 1960s in Sonneberg. (Photo P. Kroll. Courtesy of P. Kroll)

---

[10] Meinunger (1979).

[11] Weber (2006).

The Last Year

In August 1967, the new vice director of the Institute of Stellar Physics, Jackisch, started to travel from Sonneberg to Prague in the only company car of the observatory, while his old, already visibly marked former boss with his wife and staff set off from the railway station to the General Assembly of the International Astronomical Union. Hoffmeister was monitored all the time in Prague by other astronomers named IM "Sonne" and "IM Stern"[12] working as informers for the MfS. They also reported that Zimmermann had *very close contact with Dr. Weigert, who had left Jena and the GDR and was now working in Göttingen.* At the reception and the farewell party, it was noted *that Dr. Z had always stayed close to Prof. Kippenhahn from Göttingen.* Starting with the conference in Prague until the end of the GDR, the MfS recruited willing and informative astronomers from all institutes to secretly observe other astronomers. As a result, the MfS demanded that the authorities at the Academy take measures *to improve preparation for international events in other socialist countries and to create a guarantee that scientists such as Prof. Hoffmeister from Sonneberg would no longer be sent to such events in the future.* From the beginning on the secret service and its agents wrote their reports in order to suppress suspicious-looking persons in science.

All directions were now closed to the aged astronomer and the Sonneberg restricted frontier zone proved to be his only remaining refuge. His posthumously published conference report describes the event in the pre-revolutionary Prague with references to Tycho, Kepler and the devastation caused by the 30-years' War—and also the meeting of the Commission for Variable Stars in great detail. Then he managed to write the preface to his book, which he started too late. The article in Astronomische Nachrichten—submitted on 11 January 1968—on 223 new variable stars was compiled by Hoffmeister shortly before his death. "Unfortunately, he was no longer able to write an introduction."[13] He died in Sonneberg on 2 January 1968 after a stay at Jena University Hospital. On behalf of the Astronomische Gesellschaft, Zimmermann laid a wreath at the grave; Hoffmeister had been a member since 1917 and vice president from 1954 to 1960.[14] A gracious fate saved him from the coming political impertinences, the tanks in Prague, the forced exit of all GDR astronomers from the Astronomische Gesellschaft and the coming academy reform which will bring the total dominance of the state party SED in science.

---

[12] IM = Inofficial (invisible) employee of the MfS.

[13] Hoffmeister (1968).

[14] Kippenhahn (1968), Dick & Ackermann (2025), p. 247.

6

Lauter vs. Undercover

Drive Hunts

On 18 September 1972, a special commando of the MfS secretly searched the villa of Professor Lauter in Kühlungsborn (Fig. 6.1). A radio (*suitable for short-waves*) and a camera were photographed, self-composed crossword puzzles, the existence of which had been pointed out by an informer, were copied. Another informer had noticed suspicious columns of numbers on a small sheet of paper in Lauter's Berlin office. The suspicion of espionage by an

Fig. 6.1 Ernst August Lauter (1920–1984), General Secretary and Vice President of the Academy of Sciences in the GDR 1968–1972. (Courtesy of G. Entzian)

© The Author(s), under exclusive license to Springer Nature Switzerland AG 2026

G. Rüdiger, *The Astrophysical Observatory Potsdam - Triumph and Tragedies*, Astronomers' Universe, https://doi.org/10.1007/978-3-032-06294-9_6

officer group of the MfS was thus considered to be substantiated, and an operative case was opened for treason and corruption. There had been indications that Lauter was *utilising the potential of the Academy of Sciences on topics recommended by Western countries.* Because of him, the USA and West Germany were promoting *cooperation with the GDR in the field of space research*, which was not compatible with the Soviet INTERKOSMOS programme. At the beginning of 1972, the GDR leaders were informed that the General Secretary (GS) and Vice President of the Academy of Science, Lauter, prevented *the concentration of resources on INTERKOSMOS* because of his negative opinion about the cooperation with the Soviet Union and its INTERKOSMOS programme; he preferred *international projects with significant participation by the USA.*

As a consequence, a few weeks later Lauter lost his top position in the Academy but returned to work as Director of the Institute for Solar-Terrestrial Physics, which he had already headed from 1966 to 1969. With numerous incorporations[1] in 1969, he had formed a large institute for solar-terrestrial physics—pushing an international trend—in order to investigate the influence of the variable sun on the Earth's magnetosphere, its air envelope and its inner currents, the measured temperatures and the global climate.

Lauter received his doctorate from the University of Rostock in 1950, the following year, at the age of 31, he became Director of the newly established Observatory for Atmospheric Research in Kühlungsborn and also Professor of Atmospheric Physics in Rostock. As General Secretary of the Academy he had to supervise a massive internal reform. He had travelled to Sonneberg himself to replace the elderly but lively director Hoffmeister without any plan for the day after. Instead he believed that the new generation of directors, like him, wanted to establish an effective science system that would be able to modernise the entire state. He couldn't imagine how vulnerable a scientific system is that systematically replaces output with attitude. "You have to be the first, even if you still lack perfection" was his slogan in doctoral seminars which certainly was too much for almost all SED members of the Academy. In June 1971, the IM "Astronom" had denied almost all leadership qualities of Lauter to his MfS officer. Lauter would always highlight only his own results and demands *that weak scientists should be gradually excluded* from the Academy. In fact, Lauter did not know what to do with "weak scientists"—whether or not state party members—and this will have worried the informer greatly.

---

[1] Einstein Tower Potsdam, Satellite Station Neustrelitz, Test Station Juliusruh, Observatory for Ionospheric Research Kühlungsborn, Observatory for Solar Radio Astronomy Tremsdorf, Observatory for Geomagnetism Niemegk.

The new leaders appointed after the academy reform had no scruple of contact with the security officers, sometimes turning the meetings into therapy sessions for their hidden battles. The numerous reports of the informers brought the occasionally choleric Lauter into ever new difficulties. Lauter's deputy as General Secretary, Heinz Stiller, complained that he had *repeatedly warned leaders of the Academy... against appointments of the current General Secretary to higher functions.*[2] In the spring of 1972, the director of the Meteorology Service, Wolfgang Böhme, informed the MfS that Lauter was *focussing on the further connection between COSPAR and Interkosmos* against the instructions of the GDR government. This was enough for the secret service to use its sharpest weapon, namely to pull the card of conspiracy and treason. In a special briefing, several powerful SED leaders were informed that the General Secretary and Vice President of the Academy did not take a "clear position on the SED and state leadership" and rejected the "separation from the FRG". *Characteristic of Prof. L.'s attitude to the SED and state leadership was his behaviour on the occasion of the visit of Erich Honecker to the Central Institute for Solar-Terrestrial Physics (ZISTP) of the DAW on 2 February 1971. Prof. Lauter decided in his personal environment and in management meetings not to evaluate this visit as significant.*[3] Only two decades earlier, such statements had cost him the entire existence; now, in June 1972, Lauter lost his top position as General Secretary of the simultaneously renamed Academy of Sciences of the GDR (AdW) but this was the first step only.

Right at the beginning of the space age, shortly after the launch of the first Russian satellite "Sputnik", the Committee on Space Research (COSPAR) was founded in Paris as a global forum for research the Earth's atmosphere. Lauter soon became chairman of a working group, and from 1967 on he was a well-neworked office member of the committee. COSPAR aimed for maximal transparency; everyone should be able to submit scientific data and everyone should receive all the data needed. The INTERKOSMOS programme, which was launched much later to integrate non-Soviet technology into Soviet space missions, demanded strict secrecy on the order of MfS-chief Mielke. INTERKOSMOS employees had to be confirmed according to political criteria, and many of them were eliminated. Those who, like Lauter, were mainly concerned with science, walked on thin ice. Lauter demands, according to one informer, that *science must be determined by the scientists and not by officers.* In autumn 1970 he is said to have rejected Stiller's appointment as an Academy member, which was immediately reported to the MfS and must have

---

[2] Buthmann (2020), p. 746.
[3] BArch, MfS, ZAIG 1974, Bl 1–13.

motivated them to intervene in favour of their hidden colleague. Almost all of Lauter's science policy formulations later became generally accepted overnight and could all have been on the banners of demonstrators if they had existed in GDR academies and universities.

The Finishing Shot

Lauter wanted to explore the Earth's atmosphere as a whole, using ground-based radio waves, balloons and rockets from his many observatories and from space. The sun determines the state of the high atmosphere with its variable magnetic activity. He was convinced that this concept also required control over solar research—to this end he had placed the Einstein Tower, the Tremsdorf radio observatory and the Niemegk magnetic observatory under his control. The return of these observatories to the academy institutes on the Telegraphenberg later in 1984 formally marked the failure of Lauters concept of solar-terrestrial physics on a large scale. The INTERKOSMOS programme not only promised no progress at all in this respect. The Russians should pay if they want something, we are not the boys of the others, he is said to have rumbled, the purpose of cooperation is the solution of physical problems of the high atmosphere. For years, an ambitious quartet of officers from the MfS investigated Lauter for *actions against the national economy* with the threat of prison of up to 5 years. They included a bunch of informers from his closer circle, "Hans", "Bernhard", "Weiß", "Pavel", "Licht" and "Marianne". In addition, there were multiple break-ins into his living and office rooms, checking his bank account, medication and the properties of his pens. A single white sheet of A4 paper, allegedly suitable for secret writing, was monitored almost monthly for changes in the location in the office. He was classified as an *enemy base in the security area of space research, constantly propagating its international character.* Böhme alias "Hans" informed his personal officer that Lauter had proudly reported to him that the British and the Americans had pushed through his election as the person responsible for the high atmosphere at COSPAR, and that his own country had not even nominated him.

A few years earlier, Georg Dautcourt from Sternwarte Babelsberg had experienced a very similar situation when he was elected as the GDR representative at the Copenhagen Conference on Gravitation and Relativity Theory in July 1971. Dautcourt vehemently disagreed with the Academy's command for his immediate resignation.[4] Also, the political aggressive attitude of the GDR

---

[4] Dick & Ackermann (2025), p. 391.

against Israel played a role in this battle, because the gravitation conference wanted to organise their next meeting in Israel against the massive opposition of the Eastern bloc. Only the local personal committee of the ZIAP prevented Dautcourt from being dismissed from the institute. Lauter, as the General Secretary of the Academy, also refused to accept drastic actions. As a result of the disputes, for many years Dautcourt's numerous science results were only be allowed as published in the collapsing series of Publications of the Astrophysical Observatory and the Monthly Reports of the AdW, and he himself was placed under the disciplinary control of the managing director Ruben. Requests for regular reports, however, were consistently ignored by Dautcourt. In March 1976, Ruben brusquely rejected Dautcourt's application to do relativistic astrophysics with a small group in Babelsberg, but already in June the solution of all problems was promised if and only if he, Dautcourt, would provide periodic reports about the personal affairs in the Sternwarte. Dautcourt refused. When handing over the regular Christmas bonus, Ruben wished him "continued good scientific success", which Dautcourt coolly acknowledged with "Thank you, too".

The MfS officer group required that their minister may also remove Lauter as director and withdraw him from all international projects. Minister Mielke indeed gave his authorisation at the beginning of 1974. Then the new General Secretary Grote explained to his precursor the travel ban imposed on him. Lauter fought for his ideas: INTERKOSMOS is only equipment, whereas COSPAR is science, we have to be first on a global scale, and the solar-terrestrial physics is the most important research because environmental policy depends on the results. In the end he became desperate: I do not know where the conspirators are, I have always practised discipline. What is the case against me, Lauter asks. Nothing, as far as I know, Grote answers, nothing at all, unless you will now provoke an audible protest, this would be the end. The personal letters of resignation to prominent recipients all over the world that were forced on him, were indeed checked for secret messages before they were sent, mostly by "Hans". The Minister for Science and Technology to whom Lauter had once denied a desired professorship at the Academy, triumphed that *the only question to be decided was INTERKOSMOS or COSPAR, which had nothing to do with your personality,* he said, against his better knowledge, to Lauter's face.[5]

Lauter recovered from the punishment, still he was head of a large, well-organised research institute. Lauter was *still doing what he had always done*, reported an informer, in Juliusruh he built a huge antenna network to

---

[5] Buthmann (2020).

investigate the ionospheric wind systems. He was pursuing a spectacular approach at the time: the sun shapes the earth's atmosphere through its variable radiation and magnetic fields and this determines the human environment. Such planetary environmental research had not existed before on this scale anywhere in the world and even similar programmes were still lacking internationally. On the occasion of a first lecture in Potsdam he emphasised the fairly precise knowledge of the physical parameters of the atmosphere, in contrast to the chemical and biological parameters, about which almost nothing was known. The audience was interested, and Stiller as chairman of the meeting, promised further discussions on this new topic. In a representative anthology, Lauter formulated as early as 1975, the importance of research into near-Earth space as part of future environmental research. How fluctuating is the irradiance from the sun, how does the "thermostat" that regulates the temperature of the biosphere work, how does the geomagnetic protective shield function? The search for answers is "not just a mere refinement of our knowledge, but a fundamental social necessity".[6] Similar concepts will be developed decades later, and the discovery of the role of the carbon dioxide and methane in the heat balance of the Earth atmosphere would certainly not have escaped him when he would have been allowed to work. In the same publication, the future Vice President of the AdW, Stiller, explained the distribution of mass and angular momentum in the solar system as "confirmation of the basic postulates of our Marxist-Leninist philosophy", but with a view to Lauter statements on the development of the biosphere as too speculative.

The four MfS officers were once again alarmed, they now wanted to make their own science politics by turning big wheels. Hidden informers who worked as distinguished scientists and enjoyed Lauter's confidence reported that he was working *in order to achieve a spectacular scientific success that would restore his international reputation.* After he had also presented his theses on environmental research[7] in the physics class at the AdW—successfully, as he believed—a final scandal in Potsdam has been organized. R. Rompe chaired a meeting at which Lauter was to prove his thesis, that all societies must be able to survive global climate catastrophes. Rompe asked, instead to understand the nature of the matter, what this permanent *pessimism about the weather* shall bring. *Catastrophe theories are born of the capitalist system,* said the Einstein expert Treder *to divert attention from the crisis of capitalism.* Director Böhme

---

[6] Lauter (1975b).

[7] Lauter (1975a).

reported to his MfS officer that only the attitude of the people from the Academy-Institute for Science Theory was destructive who believed everything and took it seriously.

Lauter will have known that it was over for him. He declared after the meeting that he would cease these activities and also withdraw his new institute concept. President Klare immediately requested the dismissal of him from the popular GDR research council. The most promising large-scale project in the years after the academy reform was destroyed by paranoia. Now everything happened very quickly. The current leader of a regional district administration had been as a friend of the Lauter family for many decades. Lauter lost all distance from his provincial friend: *The Soviet Union hasn't been able to build intelligent satellites for decades, they can't do it. The Americans had made a decisive contribution to research with the moon landing. They are now leaders in science and his Kühlungsborn observatory has also become a top address worldwide.*[8] The long-standing family friend immediately informed the MfS about Lauter's statements, Lauter had full confidence in him. After years of surveillance the officers had finally found what they were looking for. According to their report Lauter stands for the rejection of the SED leadership of science in the GDR, discredits the successes of the Soviet Union, admires the USA and hopes to participate in NASA projects. The special division of the MfS responsible for criminal matters had fully agreed with these conclusions, but unfortunately the names of the oversea agents were missing. *The available evidence is not sufficient.* More precisely: apart from baskets full of slander, the officers had nothing to present. Nevertheless, in August 1976, Stiller informed Lauter that he would no longer be proposed as Director of his institute. The INTERKOSMOS co-operation had entered a new phase, for which he had not been confirmed. Lauter was so affected that he was not able to make any statement.

Lauter retired to his home institute in Kühlungsborn (Fig. 6.2). He worked tirelessly—without any official function—on ionospheric physics, interplanetary magnetic fields and solar-terrestrial relationships, often with younger colleagues in his nearby house, never in his institute office. When asked what had happened to him, he answered truthfully that he did not know. In his last year, the Academy published an article by him in an issue entitled Development Trends of the Earth as a Planet on the current directions in his research: The oldest observed changes in nature include excessive weather patterns, luminosity phenomena in the atmosphere and variations in the spot coverage of

---

[8] Buthmann (2020).

Fig. 6.2 50-m-antenna in Kühlungsborn 1957 of the 33 MHz radar systems for locating auroras and meteor trails with villa of the institute administration. Entzian. (Courtesy of G. Entzian)

the sun. This strongly remembers Förster's original concept for the AOP on the Telegraphenberg. Lauter digitised the magnetic observations and found that "significant correlations with the temperature of Central Europe can only be established for the winters up to the end of the last century". Since the pronounced maritime climate period at the beginning of this century, he reported, such a correlation of climate with solar activity is no longer recognisable and "one may suspect that industrial interventions in Europe's temperature regime (CO_2 enrichment) already dominate in the warm winters of the 1970s".[9] The overstrained academy administration hid the result behind a "for official use only" stamp. Four years later, in 1988, the IPCC[10] was founded by the United Nations in New York to answer Lauter's questions on a larger scale. The famous hockey stick diagram only dates back to 2001. Lauter died in 1984 at the age of 63 in the University Hospital in Rostock, not without

---

[9] Lauter (1984).

[10] IPCC = Intergovernmental Panel on Climate Change.

having ensured before that no one from Berlin and/or Potsdam would be allowed to speak at his funeral service. The previous year, he had said in a television broadcast that socialism had sofar been built in the GDR under good climatic conditions but nobody knows, however, what will happen in a worse climate. Don't think that everything will stay as it is, this was his final warning to the society—spoken in the language of the time—but the TV director cut it out.

Part III

Last Years of the Observatory, the Academy of Sciences and the GDR

7

Observations, Theory and North Korea

Max Steenbeck in Potsdam

In the late autumn of 1965, a heavy Tatra limousine passed the open iron gate with the lettering OBSERVATORIEN in Antiqua on the Telegraphenberg near Potsdam. The gate keeper had left his post and greeted the car almost militarily, because none other than Max Steenbeck, newly appointed Chairman of the GDR research council, member of the Academy Presidium and Institute Director in Jena, hurried to get to the Astrophysical Observatory in time for a seminar (Fig. 7.1). He was late, but did not want to miss even the introductory words of the announced lecture on Magnetic Stars by Lore

Fig. 7.1 Max Steenbeck (1904–1981). (Courtesy of Förderverein Großer Refraktor Potsdam)

© The Author(s), under exclusive license to Springer Nature Switzerland AG 2026
G. Rüdiger, *The Astrophysical Observatory Potsdam - Triumph and Tragedies*, Astronomers' Universe, https://doi.org/10.1007/978-3-032-06294-9_7

Oetken. The driver rushed to the top of the mountain and to the "Kaiser-Eingang" of the Great Refractor building opposite the main building of the AOP. Wempe awaited the guest at the open door, who walked wordlessly to the first row in the semi-circular, half-filled seminar room, lit a cigarette and the lecture could begin. Oetken: "It has now become increasingly clear that magnetic fields are not a random phenomenon, but a generally expected property of cosmic matter. Many cosmic processes cannot be understood if the effect of magnetic fields is ignored."[1] Steenbeck at his Institute for Magnetohydrodynamics had just answered the question how the fields can be maintained as to whether turbulence in conductive liquids can do more than destroy magnetic fields in a short time as visible in sunspots. The answer is that—if the turbulence is inhomogeneous and rotates fast enough—magnetic fields can be self-excited by a dynamo process. He believed that this could be the long-sought explanation for the widespread occurrence of magnetic fields in the universe: the planets, sun, stars and even galaxies carry turbulent flows within them and they rotate albeit at very different speeds. He[2] had already initiated this idea in Jena in 1960: In a conductive rotating homogeneous turbulence field, he had diagnosed an electromotive force that points perpendicular to the rotation and the electric current. Compared to the known decay without rotation, the decay of the magnetic fields is significantly slowed down but not fully stopped; this construction—in Steenbeck's language—only worked as a motor and not as a generator, its design was still too simple. It was later shown by K.-H. Rädler that the generator indeed had worked after adding differential rotation.

Steenbeck had also recently appeared at the all-German AG-conference in Eisenach in 1965 with the announcement that the magnetic puzzle had been solved, reading word for word with his poor eyesight in the semi-dark room, while his assistants had to write on the blackboard or direct the slide projector. Almost 200 astronomers from both East and West Germany, among them established researchers such as Kippenhahn, Weigert, Mezger and Schmidt followed the performance of the Jena trio with amusement, but were also impressed by how chaotic flow patterns can be so elegantly treated mathematically.[3] The small-scale turbulence elements were simply averaged out so that only the behaviour of the remaining large-scale magnetic fields must be calculated. Typical for the new theory is that, in addition to the visible external magnetic fields of the celestial body, there must always exist internal fields of

---

[1] Oetken (1966).
[2] Steenbeck (1961).
[3] Steenbeck & Krause (1965).

at least the same strength, which are invisible and perpendicular to the external fields. After the presentation, there was silence, no questions, no comments—nobody in the audience had ever heard anything like this. In a further contribution, Krause demonstrated with a few equations, that a weak meridional flow on the surface of a gas sphere causes the observed differential rotation of the sun if only "the matter at the surface flows towards the equator".[4]

Due to the lack of applause, Steenbeck returned to Jena immediately after his talk, leaving both his assistants Krause and Rädler at the meeting without any comment. Nevertheless, he further developed his ideas: Turbulence and rotation was all that is needed, rotating turbulence is indeed present everywhere in the Cosmos, how universal is thus the new approach? The Nobel Prize for Physics, he wrote to the Swedish Nobel committee at their request, should soon go to a magnetohydrodynamicist, such as Alfvén is one—or even he himself? Then Professor Alfvén should at least be among the prize winners.[5] Alfvén did indeed receive the prize soon afterwards.

Steenbeck, Krause and the doctoral student Rädler had previously published a German-language paper on dynamo theory—as the new branch of research has since been called—in a West German journal, which has been cited 675 times[6] internationally to date, a scientific milestone, possibly the only one in the history of the GDR science. Dynamo theory considers the formation of cosmic magnetic fields as self-excited: if the turbulent celestial body rotates fast enough, infinitesimal small magnetic seed fields are amplified to the observed values.

Steenbeck recruited students from the current fourth-year physics programme at the Jena University to support the development of the theory. In particular, he wanted to know whether the calculated magnetic fields could oscillate in an 11-year rhythm similar to that of the sun. At the turn of 1967/68, to his satisfaction, it could be reported[7] in his villa in Jena that they do indeed if differential rotation exists. "He's a physicist, not just a mathematician" was the highest praise one could hope to get from him.

After the Oetken seminar—nobody but Steenbeck ("Where is the turbulence in these magnetic stars?") had asked a question—Wempe explained the plans for the observatory in his formidable office: a group for spectroscopy of

---

[4] see Kippenhahn (1963). Observations show that the plasma at the solar surface flows polewards.

[5] Helmbold (2016).

[6] Steenbeck et al. (1966), data according to web of science core collection, summer 2023. The first reference to this paper in the literature was by H.K. Moffatt in 1970.

[7] By the present author.

magnetic stars with observations with the 2-m-mirror in Tautenburg[8] and a group for photometry with observations from Sonneberg. No words about the stars with magnetic cycles, although Eberhard and Schwarzschild with the Great Refractor and the Zeiss triplet[9] had discovered the magnetic relationship between the Sun and the cool giant stars Arcturus and Aldebaran. The old report for the Academy describes how in 1900 overexposed spectra were produced at Vogel's request in order to reach the weak UV spectral range. Eberhard found a sharp emission core in the K-line of Arcturus, "just as is observed in the sun in disturbed areas of its surface". Such "reversals" in the calcium lines above magnetically active regions had been known to solar observers. Schwarzschild had combined this surprising finding with his own observations. Aldebaran showed a similar effect, Betelgeuse, α Cass and the diffuse daylight as a proxy of the integrated sunlight did not. Particularly conspicuous emissions were found in σ Gem, which is known today as an active close binary with large starspots. The strong calcium emission was also later confirmed for Aldebaran[10] with its very long rotation period of 643 days. "The emission lines of the stars will have to be checked for any variability of their intensity in analogy to the sunspot period", they had written in 1913, an astronomical sensation which, however, has been forgotten worldwide until 1957; the first reference from Potsdam to this publication is even younger. Schwarzschild died of an autoimmune disease in May 1916 at the age of 42, having left the Russian front in March. In his final weeks, he had discussed quantum theory with Sommerfeld and gravitation waves with Einstein. "In Potsdam he was not really able to put down roots, however his work as director of the Astrophysical Observatory was fruitful and however great the recognition he received as a scholar and as a person from Berlin circles", writes Runge in his very soon obituary.[11] Why the Principal Observer Eberhard did not continue his most important discovery, we do not know, was it opposition to Schwarzschild, did he not believe in his vision? Calcium activity in stars never again played a role as an observational task in Potsdam, nobody had remembered it, not even in the tributes of Wempe and his colleagues[12] to the centenary of the Astrophysical Observatory. A dedicated observation program would have provided the Potsdam groups with useful data even under the local weather conditions. The modern calcium campaign did not start on

---

[8] First publications on α^2CVn by Oetken et al. (1970) and on 53 Cam by Scholz (1971).

[9] Eberhard & Schwarzschild (1913). Triplet: 150 mm aperture, 1500 mm focal length, from 1908 in the east dome. Great Refractor: 80-cm-objective, spectrograph III.

[10] Kelch et al. (1978).

[11] Runge (1916).

[12] Wempe (1975), E.-A. Gußmann and G. Scholz in the same issue.

Telegraphenberg, but at the Mt Wilson Observatory in Pasadena with the Hooker telescope, which was three decades the largest telescope worldwide, under fantastically clear sky. After 15 years, the results led to an earthquake in stellar physics.

Wempe also spoke to Steenbeck about the need to solve theoretical problems of magnetic stars: why are the fields usually oblique like on Earth, why do magnetic stars rotate more slowly than normal stars, why do only some A-type stars have magnetic fields? A theorist from Jena would be most welcome. Why not two, asked Steenbeck. He planned his own future without science, he will lock his own institute by himself soon after his official retirement at the age of 65, there is thus no future for the employees in Jena. After his retirement in 1969 the Institute for Magnetohydrodynamics of the AdW has indeed been closed. He had no longer interest on its further existence and he preferred to focus his activities to campaign for nuclear disarmament in East-Berlin.[13] He is seemingly still powerful, but he is not a party member, occasionally the SED-leaders and the ministers are terrified at his ideas, sometimes his contributions as head of the GDR research council are not even published. He was aware of the intentions of the Academy to appoint only loyal members of the state party with Moscow experiences to head institutes, preferably speaking Russian; and he was still convinced that the theory of cosmic magnetic fields was finally worthy of a Nobel Prize.

The former Siemens director Steenbeck was one of the most active returnees from Russia who, like Ardenne, Barwich and Hartmann, conducted military nuclear research in Suchumi or, like Albring, Hoppe and Erich Apel had continued to develop the Peenemunde A4 missiles on the island of Gorodomlia. After their return around 1955, if they decided to stay in the GDR, they received private contracts for prominent positions and quickly organised their entry into nonmilitary atomic energy. With Nobel prize winner Gustav Hertz on the top, they wanted to build nuclear power plants and even civil jet planes—until everything suddenly collapsed because the SED leaders stopped all plans (Fig. 7.2). The newly established faculties for nuclear physics and aircraft construction in Dresden were closed, and the Baade 152 jet plane even crashed in 1959. Barwich fled to the USA and Apel as head of the central plan commission—the heart of the socialistic economy—shot himself in his own office. Steenbeck after all this chaos had been given a comfortable, personal asylum in the Institute for Magnetohydrodynamics in Jena from 1959 until his official retirement.

---

[13] Personal communication Klaus Steenbeck (son).

Fig. 7.2 Rheinsberg nuclear power station, whose construction was decided in 1956, after the world's first nuclear power plant had just gone into operation. Construction of the second unit, initiated by Steenbeck was never started. He headed the reactor construction in East-Berlin from 1957 to 1960. (Courtesy of EWN GmbH)

Steenbeck was now looking for reliable housing for his most promising employees, absolutely in his neighbourhood; soon he would be living in East-Berlin forever. He planned a big dynamo experiment, writing letter after letter on the subject as a scientist who would very much welcome the realisation of such a fundamental experiment in the socialist world, because "otherwise it would surely be carried out in the USA or France in a few years".[14] He wanted to build a huge sodium circuit and would need a container with around 10 m^3 of liquid sodium and a pump that could circulate this amount in about one second. Indeed, a rotating sphere with a diameter of 3 m filled with 14 m^3 liquid sodium has later been in operation at the University of Maryland for several years. His early numbers base on the "α-box experiment", which ran until 1967 at the Riga Physics Institute, in which the existence of the dynamo effect was demonstrated in the laboratory with flow speeds of several m/s and magnetic field strengths of 1000 Gauss. These dimensions were too gigantic for the GDR officials, they sent him with his ideas directly to Moscow. The Ministry for State Security had already learnt in August 1966 that on the occasion of a visit of Steenbeck to Soviet Academy President Keldysch, it had

---

[14] Archive Stefani.

been decided *that the Soviet Union would carry out all major experiments for Steenbeck on the subject of researching the Earth's magnetic field. Prof St. was very enthusiastic about this decision.*[15] Steenbeck had already announced the intention to operate a first dynamo experiment in Soviet Physics Doklady in (1968).

Wempe refused the personnel issues, saying that there was no suitable living space in Potsdam, that the city was built on a swamp and that there was no chance of success. The real reason for the deficit was that 9000 of the 37,000 flats in Potsdam were used by the Soviet army. Only an undemanding young candidate from Jena could just about be accommodated without family. The disappointed Steenbeck promised to train an undergraduate student from the Sternwarte of the University of Jena in dynamo theory for Potsdam. On the journey back, he was thinking more about the magnetism of the Earth than that of the stars. He had the names of all experts of geomagnetism from Telegraphenberg written down so that he could order them to Jena shortly. This invited presentation took place in the spring of 1968 with lectures by Steenbeck, Krause, Helmis and Rädler. One or two experts from Potsdam announced that they already had similar ideas but they were simply too busy for own publications.

In September 1968, the promised new graduated student will arrive at Telegraphenberg without family and the Jena employees Krause and Rädler will start their relations to Potsdam with scientific contributions during the international conference of the Profs Lauter and Jäger at the 5th Consultation on Solar Physics and Hydromagnetic.

Institute for Stellar Physics

In the Sonneberg Observatory's annual report for 1966, Hoffmeister promoted the previous group leader, Jackisch, to deputy director while Wenzel remained a group leader. Despite several reminders from Berlin, Hoffmeister had never named a successor, so that at the turn of the year 1966/67 a panel named Physics-North decided this question. A new structure for astronomical research within the Academy was defined in November 1966 under the leadership of the powerful prospective General Secretary Lauter. In the future, there were to be only three main fields of astronomy, namely solar-terrestrial physics, stellar physics and extragalactic physics. The Babelsberg Observatory formed the third pillar in Lauter's concept of astrophysics at the Academy as the Institute for Relativistic and Extragalactic Research. Lauter's plan to focus

---

[15] Archive Helmbold.

on searching solar-terrestrial relationships was not stupid; Wilhelm Förster had already proposed in 1872 a solar-telluric institute as a concept for the Observatory to be founded in Potsdam. However, the new concept contained two conditions: The AOP was to hand over all its solar research potential to Lauter and the Sonneberg observatory its independence (Fig. 7.3). On the other hand, it came down to Wempe as the director of a future Institute for Stellar Physics; the much more productive Sonneberg Observatory could not name a rival, especially not in the academic field; no one from the Sonneberg staff except Hoffmeister had ever given lectures at a university. On 24 January 1967, the almost 75-year-old patriarch wrote to the Academy President: *I don't understand why staff changes should be kept secret, given that the staff members are also published in the Academy yearbook and all changes can be identified by comparison.* Had he—contrary to the rules—wanted to talk to his employees about his future and that of the observatory? Academician Hoffmeister, somewhat protected by his high reputation and the large distance between Sonneberg and the Academy headquarter in East-Berlin, had occasionally sent such letters on behalf of his astronomers.

Hoffmeister, who also was personally present at the Physics-North meetings, often responded in writing letters. In February, a few days before the panel made its final decision, Lauter received his counter-proposal: Potsdam as the Institute for Solar and Stellar Physics and Sonneberg as the Institute for

Fig. 7.3 Sternwarte Sonneberg, ca. 1975. (Courtesy of Astronomiemuseum Sternwarte Sonneberg)

Variable Star Research. Lauter without the Einstein Tower in Potsdam—this was a much too simple idea of the patriarch. Lauter believed that observatories such as the Einstein Tower were urgently needed for his new type of institute construction. After all the institutes' structure was fixed in a meeting on 17 February 1967.[16] It was also announced that Wempe *is prepared to fully support the scientific objectives of the new main research field of stellar physics and is also prepared to take the leading role in this field.* Shortly after Hoffmeister's 75th birthday, Lauter proceeded to Sonneberg to inform the astronomers. The meeting in the lecture room of the old main building had turned into a shocked silence. Only the elderly optician Rudolf Brandt rose to protest, but Hoffmeister remained in silence, he had no words because—as he knew— everything had already been fixed. People were going to lose their leader and Hoffmeister his observatory to Wempe whose scientific production was invisible for decades (Fig. 7.4). His last working day as director was set for 31 May

Fig. 7.4 Johann Wempe (1906–1980) by Tam Bräuer, 1970. (Courtesy of H.-J. Bräuer)

---

[16] Profs. Lauter, Hoffmeister, Treder, Lanius, Richter, Wempe, Lambrecht, Daene and Jäger.

1967; he was given a new employment contract as head of the tiny group Field Plan and Typology of Variables with a monthly salary of 800 M and an extra pension of 1000 M. In 1964, Wenzel had still attested Hoffmeister's fanatical attachment to his job. "He spent large sums of his private funds on the purchase of astronomical equipment for the observatory."

From 1 June 1967 the Institute for Stellar Physics existed under Johann Wempe, which consisted of the Astrophysical Observatory Potsdam and the Sternwarte Sonneberg, with Ruben and Jackisch as deputies and with Oetken and Wenzel as group leaders. Among 94 employees at the time were 16 members of the SED party, one of them even as a graduate of the powerful Astro Soviet in Moscow. The Einstein Tower and the Tremsdorf radio observatory with their Profs Jäger and Daene were transferred to the Heinrich Hertz Institute for Solar-Terrestrial Physics in Berlin-Adlershof with Lauter as the director of Physics of the high Atmosphere and Ionosphere and Schmelovsky as the director of Technology. The gentlemen had quickly reached an agreement, Wempe was satisfied; his new institute with the two deputies from the state party was obviously well positioned, and the specially trained graduate[17] promised by Steenbeck had been announced for next year to receive one of the cheap doctorand contracts. Wempe, on Lauter's recommendation, also hired in 1966 the radiation transport expert H. Domke, whose churchly activities had displeased one of the members of the interview. He is already closer to heaven than other astronomers, the director wiped the dangerous accusation from the table. Every Christmas, Wempe gave each of his numerous employees presents, often with subtle references; neither the Astrophysical Observatory Potsdam nor the Sternwarte Sonneberg, however, shall survive the political battles within the next two or three decades.

Wempe's concept was made outside the political system. At the SED convention in April 1967 Max Steenbeck as the head of a huge delegation of university rectors and academy presidents, announced the new political line that "from now on we accept scientific cooperation only when it does not interfere with our path to socialism in our own state. Anyone who wants to deny us the right to follow this way, cannot count on our cooperation, even if this cooperation could be useful for us." Instead of using such a prominent opportunity to name the deficits of the official science policy, to point out the delay in numerical research methods, to secure access to Western supercomputers and telescopes, he only formulated the ideas of the functionary apparatus. He could also have said that all GDR scientists shall from now on accept the dominant role of SED members and security officers at universities and

---

[17] The present author.

Fig. 7.5 The Observatory Tautenburg board 1960; left to right: Hoffmeister (Chairman from 1961), Kienle (Chairman until 1960), Görlich, Lambrecht, Richter (Director since 1961). The all-German board only existed until 1967. (Courtesy of Thüringer Landessternwarte)

academies. It was the socialistic primacy of politics over success, presented in his own words after a draft of the central SED headquarter by "Max Bolshoi" (as he was also known) equipped with a diplomatic passport and the highest possible monthly salary of 15,000 M. With his speech the downswing of science in the GDR started, the nascent hope that the construction of the Berlin Wall would not affect scientists was dashed. Did the intelligent Steenbeck not realise that he had dealt the death blow to science and socialism in his country just at an SED party convention? In the same year also the all-German status of the Observatory Tautenburg has been cancelled, which was founded in 1960 under a mixed directorate with Görlich, Hoffmeister, Kienle, Lambrecht, Wellmann and Wempe (Fig. 7.5).

At that time, the rusting Great Refractor on the Telegraphenberg was still in operation, and on clear nights Gerhard Böttger used the 50-cm-objective to take photos with long exposures of wide binaries (Fig. 7.6). In most cases, however, the nights suitable for observation were too widely separated in time. The number of annual observations—at the astrograph in the middle dome there were up to 30 additional nights—was almost less than ten. Güntzel-Lingner with the Hertzsprung method had resumed the astrometry of wide

Fig. 7.6 Refractor building in the 1970s. The outer skin of the dome was renovated 1985 by the Potsdam city office for monument protection

binaries with very large orbital periods for the purpose of determining the mass of the stars with extensive catalogues (Fig. 7.7). When in 1952 the Great Refractor returned to scientific work, a continuation of Hertzsprung's work was the main application for the approximately 70 usable nights per year. "Güli's" photographically achieved accuracies, mainly obtained with the visual 50-cm-objective were far above the visually achievable values.[18] The only theoretical use of these observations for the statistics of stellar angular

---

[18] Brosche (2014).

Fig. 7.7 Left to right: Peter Brosche ca. 1960; Ulrich Güntzel-Lingner (1914–1979). (Courtesy of P. Brosche/Förderverein Großer Refraktor Potsdam)

momentum was by P. Brosche student at the Humboldt University in East-Berlin (Fig. 7.7). According to his results, the angular momentum of binary stars exceeded the angular momentum of the solar system by far,[19] as if the rapidly rotating protostars were developing into binary stars and the slowly rotating ones into planetary systems. After Brosche's leave of the observatory—partly due to lack of accommodation—this much-promising approach to star formation was not continued. Instead of the time-consuming and unspectacular orbital measurements, why did they not follow Schwarzschild's calcium activities of fast rotating giant stars? The Refractor was shut down completely in the early 1970s. Its restoration and modernisation were not discussed again until 1981, when E. Gerth had convinced the engineers of the institute that stellar magnetic fields could be measured with the photographic objective if only the signal was conducted to the instruments in the basement of the building with fibre optics. Since the outer dome skin was restored in 1984 by the Potsdam city administration, this idea has occasionally played a role in the internal discussions of the ZIAP but without tangible results. Only as a result of the activities of the Förderverein Großer Refraktor Potsdam e. V. the Refractor has been fully functional again since 2006 but is no longer used for scientific purposes.

---

[19] Güntzel-Lingner (1955), Brosche (1962).

At the end of 1968, Steenbeck's closest assistant Krause was appointed to succeed G. Fanselau as director of the Geomagnetic Institute on Potsdam/ Telegraphenberg. Fanselau had managed to gather around him a considerable number of ambitious young scientists such as Best, Kautzleben, Möhlmann, Stiller, Treumann, Wagner and Wiegank, who will make a name for themselves in the future in one way or another. Krause came from the Jena School of Mathematics of Walter Brödel and had also studied physics under Friedrich Hund.[20] Due to political difficulties after his assistantship, he had moved to the neighbouring Steenbeck's institute which urgently needed a capable mathematician. At his introductory presentation in Potsdam, Krause presented his new turbulence-based answer to what he believed being the key question of all geomagnetic institutes world-wide: why does the Earth have a magnetic field? The reaction of the older classes of the audience to the presentation was largely negative, Fanselau and other long-established members of the institute missed the knowledge and appreciation of their own previous work. The academy reform that had started in the meantime, which consisted of the dissolution of the historically-grown independent institutes mostly headed by older academics, offered the young and active SED members the perfect opportunity to get rid of uncomfortable newcomers. By presidential decree of 23 January 1969, the Geomagnetic Institute was suddenly closed, the two MHD experts from Steenbeck's institute, Krause and Rädler, were sent away to the neighbouring Astrophysical Observatory, where they could form a new theoretical group with Krause as its head.

Wempe was highly enjoyed by this development and he quickly organized suitable office space. The housing issues, which would have been beyond his influence, were solved by both the new-born astronomers themselves through private flat swap deals. The establishment of the new theory group at the Observatory will prove to be a great luck. During the evaluation by the all-German Wissenschaftsrat (WR) in 1991, one of its members described the current state of the Central Institute for Physics of the Earth (ZIPE), to which the former Geomagnetic Institute had been added, as "the necessary result of a monopoly position with isolation from the outside world and at the same time a lack of leadership of the management".[21] In contrast, the same commission will emphasise the "great achievement of the dynamo theory group" when assessing the astrophysics institute.

---

[20] Appendix 10.

[21] Boch (2008).

Arrival in the Morning Haze

My last summer holidays as a student ended abruptly on 21 August 1968 when early in the morning the campground near the Baltic Sea was suddenly evacuated. It was the day the troops of the socialistic countries invaded Prague. For my diploma examination in Jena, Professor Lambrecht had wished the dynamo theory of Spörer's law explained to him, which I had formulated in my thesis at Steenbeck's Institute. As usual, Lambrecht's two-hour semester lecture on solar physics was cancelled after a few weeks; the assistants rumoured that a continuation never existed, because of too many details, "the sun was too close for real astronomers".

After entering on 1 September, the tiny loam construction with oven in the attic of an old and large house in Potsdam—certainly used as a soldier's chamber in the past—I looked for the library building of the Astrophysical Observatory at 8 o'clock the next morning. The only person I met there was the librarian; all but one of the scientists' offices remained empty that day. In Jena and Sonneberg or even in Steenbeck's MHD institute had always been activities in the mornings. I had traditionally spent the winter holidays in the Sternwarte Sonneberg, skiing one last time with the whole staff through the rather dense forest, the same route I had taken years ago on foot in the pitch darkness, all the way to the upper town, without a light and without a pass for the closed border area.

Wempe did not receive me until late in the evening and explained the evacuated state of the theory group: Ruben was currently in Moscow and Domke in Leningrad (St. Petersburg). He suggested to publish my diploma thesis on dynamo models of solar-type stars with the first oscillating mean-field dynamo, which he had heard about, in Astronomische Nachrichten edited by him. But I wanted to get started, tackle new things, not write down old ones, and ignored his good advice. Later Krause had suggested a dissertation topic on the differential solar rotation to me immediately after his arrival at the Geomagnetic Institute and, after he moved to the Observatory, I spontaneously switched to his group, ignoring the plan to work in Ruben's unspectacular theory group.

Every square metre of Potsdam's city centre is steeped in history and anecdotes, but encountering Potsdam in the 1970s was like suffering a culture shock. Broad roads with crumbling authentic baroque houses, the just blasted stones of a famous church building lying on the ground, an old city canal stinking, the loose gutters rattling in decaying streets—nobody thought they were responsible, not even the residents. Later, this socialistic concept of destroying traditional city centres was called "ruining without weapons". It hardly needed any dynamite, just rain; often one could still see bullet holes in the plaster from the WWII more than 20 years ago. The corner houses of the blocks became ruins first, only a few windows towards the street lit by naked lights in the late evenings (Fig. 7.8).

Soon after the first year of my three-year doctoral term, just months after his arrival from Moscow, vice director Ruben had already personally registered me with the MfS and promised the security officer that *R. only receives enough information to enable him to do his job. He will learn practically nothing about the Institute for Astrophysics, the research structure and the results.*[22] This was the exact opposite of all my expectations. Having just left the university, I was basically interested in all activities of my new institute. Perhaps the officer taking the report was even wondering how scientific institutions are supposed to work under conspiratorial rules. At the time, the ideological struggle of the SED focussed on the warning against the non-Marxist New Left, which does not recognise the leading role of the "working class"; perhaps the Astrophysical

Fig. 7.8 Baroque Street in Potsdam 1989. (Courtesy of Robert-Havemann-Gesellschaft/ Siegbert Schefke)

---

[22] MfS text of 27 January 1970, hand file.

Observatory at the Telegraphenberg needed special protection against such ideological infiltrations from Jena. The future development has shown that the astronomers in Potsdam and Babelsberg were indeed treated with particular suspicion by the state security and, as will be shown, with some reason.

Ruben alias IM "Astronom" had first appeared in the MfS files when he monitored the two astronomers Kippenhahn and Weigert from Göttingen during the IAU General Assembly in Prague in August 1967. "Astronom" had reported that Kippenhahn's *eyes had blinked* when Mrs Massevich had held out the prospect of closer co-operation between the groups in Moscow and Göttingen. Kippenhahn later: Shouldn't I be happy about that? In the MfS report, it was written as *if I had been happy that we could spy on Soviet astronomy*.[23] This was indeed the strange suspicion of the security people and whether the Astronomische Gesellschaft, together with the West German government, did not want to realise the idea of a still existing all-German science to be able to influence the appointments of astronomical managers. Always, the tireless IM "Astronom" watched everything and everybody: *On the afternoon of 30 August 1967, XXXX was visited in the conference room in Prague by XXXX from West-Berlin, who had worked at the academy institutes in Potsdam until 1961, and both then left the room. XXXX was no longer present for further discussions that day*.[24] Such observations in an astronomy meeting were considered so relevant that in December 1967 a secret operation (OV) "Horoscope" was launched *on suspicion of hostile activity by the Astronomische Gesellschaft and its chairman, Prof XXXX*. Why were the security agencies so interested in monitoring GDR astronomers so intensively? Possibly in contrast to other sciences, hardly any East German astronomers except the science functionaries believed that their research was only national rather than international and this was known in the administration. W. Mattig and E.H. Schröter had thus stayed home because of matters of their safety (Fig. 7.9).

One of the soon visible consequences of Steenbeck's party convention speech was the forced resignation of the East German astronomers from the over one-hundred-year-old Astronomische Gesellschaft in summer of 1969. Since Zach's first astronomy colloquium 1798 in Gotha astronomers have always wanted to know each other personally in order to talk about the results of their nightly and often isolated work. The annual meetings with welcome evening and public lecture formed the successful frame; and they were also proud of the spiritual continuity with former members such as Abbé, Schwarzschild and Einstein.

---

[23] Kippenhahn (1999).

[24] Lore Oetken and Marlene Mädlow, see Kippenhahn (2001). Names darkened by the BStU.

Fig. 7.9 Egon Horst Schröter (1928–2002) and Wolfgang Mattig (1927–2018). Freiburg 1987. (Archive Kiepenheuer Institute Solar Physics. Courtesy of W. Schmidt)

During the AG-meeting 1965 in Eisenach, Lambrecht was re-elected as vice chairman and also Jäger (Potsdam) and Zimmermann (Jena) joined the board. Over the course of the following year, Kippenhahn wanted to establish Helmut Zimmermann as his deputy, a huge gain in prestige for the Jena astronomy school. Zimmermann wrote to Kippenhahn in October 1968: "I was surprised, but also shocked, when I read that you had appointed me as your deputy. Dear Mr. Kippenhahn, please understand me correctly, first of all I am unquestionably grateful that you have placed so much trust in me.... Nevertheless, I would like to ask you most sincerely to renounce this election. I cannot accept it." The reason was that every registered astronomer from the Academy had to send a prescribed text to Göttingen or Heidelberg to declare their resignation: "Since the current policy of the government of the Federal Republic of Germany is incompatible with the political goals and endeavours of the GDR, I, as a citizen of this state, can no longer be a member of a society that has its headquarters in West Germany and thus consciously or unconsciously supports the policy of the Federal Government, in particular its claim to sole representation"[25]—formulated almost word for word with Steenbeck's upsetting text. The people from the Jena observatory were allowed to use

---

[25] Pfau & Schielicke (2013).

more personal formulations. On 22 July 1970, Zimmermann received confirmation at his private address that the "Board of the Astronomische Gesellschaft had regretfully taken note of your resignation".

There had also been a prelude to this: In August 1968, the MfS reported to the certainly amazed party chief Honecker that the (West) German Astronomische Gesellschaft *would do everything in its power to achieve closer co-operation, especially among young GDR astronomers... The AG presidency would like to adopt 'useful' young and talented scientists who are members of the SED because of their future chances to take up leading positions in the academy institutes.*[26] Schöneich (Potsdam) and Wenzel (Sonneberg) were both mentioned by name in Kippenhahn's concept paper. A challenging idea, but East-Berlin was not so easily duped in this area. Kippenhahn's plan, conceived during or shortly after the Eisenach AG meeting, had the first consequence that he himself came into the focus of the MfS machine. Whether they liked him – it was the heyday of the '68s in Göttingen – or feared him remained unclear. In any case, a contact person who was part of the GDR delegation to the IAU meeting was hired in Jena to skim as much information as possible from the conference participant Kippenhahn *without being syphoned off himself.* According to the MfS record there was allegedly only a 5-minute conversation in Prague with Kippenhahn and Weigert– the contact person was the young Fritz Krause from the Institute for Magnetohydrodynamics who would not be able to free himself from this connection for many decades. Ruben described to his officer in 1973 how, at a meeting in Warsaw, he had explained to Kippenhahn, who had taken the resignation process very much to heart and had seen a personal responsibility, *that the Astronomische Gesellschaft was still a national society of the FRG and that it was not customary to join purely national organisations*—the typical argumentation of SED leaders. The fact that still a few all-German professional organisations survived the socialistic revolution must have both surprised and angered the SED leaders, with the given consequences.

The forced resignation of the East German astronomers coincided with the year of the moon landing, which was broadcast on West German TV well after midnight on 20 July 1969. The Soviets had lost the race to the moon, although they had previously managed the first soft unmanned landing before the Americans. The event of the century was only discussed in closed circles at the AOP; there was no public event nor was there a private invitation to watch television together early Monday morning. Wempe and his sister had only dared to invite the Jäger family to celebrate the moon landing at night, no one

---

[26] BStU, MfS, ZAIG 1543.

else. In Jena, hardly anyone had problems with West German television and so we celebrated Armstrong's "small step for a man..." among friends of former physics students. At the workshop on interstellar matter organised by the Universitätssternwarte Jena in March 1986 in a guest house in Thuringia, the flyby of the ESA probe Giotto by Halley's comet, which was broadcast by the western TV, was of course watched together with the West German participants[27] and their whisky—inconsistently with Potsdam standards.

With the AG disaster, the political authorities' furore for isolation of GDR from the western democracies had by no means reached its peak. Up until then, it had been quite possible for astronomers to go on official trips to congresses, visits and observation campaigns in western countries, depending on the financial situation. As late as July 1953, for example, the Academy President had decided that three astronomers each from Potsdam, Babelsberg and Sonneberg could travel to Bremen for the annual meeting of the AG; the directors would only have to make the selection. Now, just 15 years later, the party functionaries were saying that they alone decide on contacts and all foreign relations, even including those with external funding! There were never any written or verbal reasonings for refusals; in the competition for networks and collaborations, the majority of scientists were excluded for all times. The 529 people with a business passport for "all countries"—later called "traveller cadres"[28] represented research at the AdW alone and without the pressure to present any own scientific result. "An entire epoch ahead", as the highest SED leader Ulbricht had unironically described this state of the country. "In one or two generations, we will match North Korea, with new insights into the integration of the exponential function e^x", Krause had whispered when a North Korean delegation visited the Astrophysical Observatory. In their Russian-language contributions, they had already solved all known relevant astronomical problems, from the expansion of the universe, the understanding of quasars, the solar 11-year cycle and the significance of planetary distances. "Some people say that we should have a telescope, it doesn't have to be very big, just to confirm our findings", mentioned the chief of the delegation at the end. All the delegation members showed their party membership with a badge with the face of the actual political leader on the jackets and the chief was simply recognisable by a particularly large badge. Commentary by Ruben: Kim-Il-Sung, what an important and persistent leader.

At the new stellar physics institute, H.-J. Hubrig and E. Gerth began to modernise an existing Abbé comparator for measuring smallest distances in

---

[27] Chini, Habing, Mezger, Staude (Heidelberg).

[28] In German: "Reisekader", status 1974. AdW 1991: 16,101 employees, 7479 of whom were university graduates.

line spectra in analogue-electronic form and to equip it for future digital data processing. Relative line splittings of 10^{-6} were now measurable, corresponding to a magnetic field of 1000 Gauss, an enormous increase in accuracy. The magnetic fields of the stars 53 Cam and ε UMa could be determined in this way during all rotation phases using spectrograms from 2-m-mirror telescopes, mostly from Tautenburg. "For the decision between different models of magnetic stars", the increase in accuracy is of actual importance, the team members wrote in the Zeiss journal Jenaer Rundschau. After a meeting of the Sub-commission on Magnetic Stars of the Academies of Socialist Countries (see Fig. 7.10) at the special observatory in Zelenchuk[29] several dozen spectrograms from the 6-m-mirror telescope there were also analysed in Potsdam. The exposure times for the analysed stars were reduced from around 100 min to 10 min compared to the Tautenburg telescope. The earliest demonstrable magnetic field measurements from these spectra date from 1983 and were for the magnetic supergiant ν Cep.[30]

Another perhaps too early design by Hubrig initiated by Wempe 1969 was a mechanical light-electric scanning machine for automatic spectrum evaluation (GASPEK), with which line profiles from up to 6 spectra could be measured, digitised and stored on tapes simultaneously. The device is still available

Fig. 7.10 Sonneberg 1974, sub-commission Magnetic Stars of the Academies of Socialistic Countries; from right: Panov, Stępień, Gerth, Bartl, Gußmann, Domke, Musielok, Rüdiger, Lange, Hildebrandt (incomplete)

---

[29] Further meetings (always without Western participations) in Shemacha (1973), Sonneberg (1974), Prague/Ondrejow (1978), Szombathely (1983) and Zelenchuk (1987).
[30] Gerth et al. (1991).

Fig. 7.11 A former chicken coop with two separated rooms as the observation house for the 35-cm-reflector telescope from Sonneberg at the Shemacha Observatory in Azerbaijan. Despite all the promises, there was only one clear night in the winter of 1969/1970. (Photo W. Fürtig. Courtesy of W. Fürtig)

today, but wasn't intensively used due to inadequate computing and storage technology as well as turbulences in the institute's life.

Photometric observation of stars with magnetic fields was not seriously undertaken either in Potsdam (70-cm-mirror) or in Sonneberg (60-cm-mirror). Rather, Schöneich, a graduate of the Sternberg Institute in Moscow, preferred to set up an external station in the south Caucasus by operating a 35-cm–mirror telescope—constructed by W. Fürtig in Sonneberg—in the Shemacha Observatory in Azerbaijan (Fig. 7.11). Observations started in 1969. Carl Zeiss Jena had previously installed a 2-m-mirror telescope there, which seemed to justify the choice of the location. The question of missing local observation protocols to discuss the possibilities had often been asked but never answered. The journey via Moscow took two days, the institute staff in Shemacha was permanently at odds and there was only little infrastructure; there had never been any astronomical co-operation between the institutions. The observatory only provided water, electricity and accommodation. A twin reflector built in Potsdam with two 35-cm-mirrors was soon installed in 1972 which had to be repaired annually on site by a maintenance team from Potsdam and only reached its full capacity in 1981 shortly before the site has been closed (Fig. 7.12).

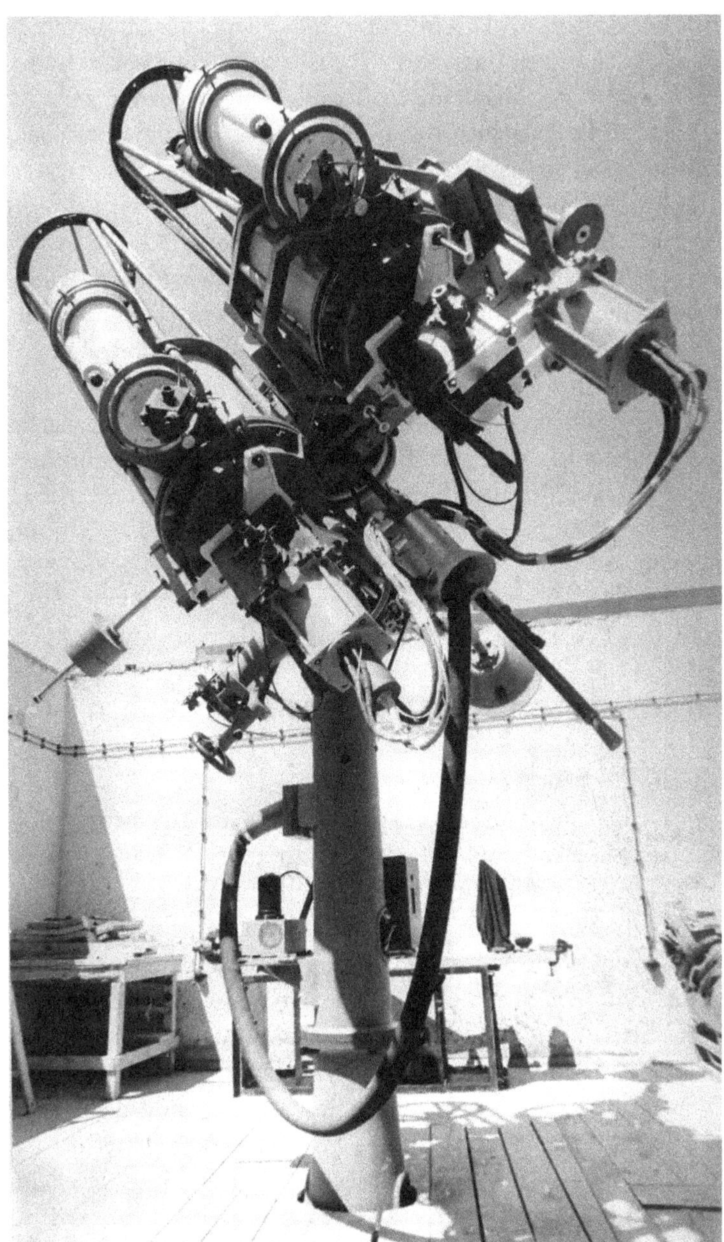

Fig. 7.12 The twin reflector in Shemacha/Azerbaijan. Design and electronics: H.-J. Hubrig. Construction: J. Czeschka, transfer 1972, lost. (Courtesy of M. Krause)

Descriptions of the Potsdam instrument in Shemacha can hardly be found today. There are only brief notes in a thin Russian-language volume printed by the Azerbaijan Academy in form of conference proceedings. Almost all of the published light curves came from the 35-cm-mirror of Sonneberg; it was used to measure rotation periods of peculiar A-type stars, periods in combination with magnetic field measurements of binary systems with an Ap component were determined together with photoelectric observations in 10 colours. In a short note in the proceedings it is reported that with the twin telescope between 1973 and 1981 only 5 of 16 Ap stars examined showed the desired weak nonperiodic light variations. Nevertheless, it was reported that the publications linked to this telescope "were an essential part of the successes of the multilateral cooperation". The observation campaigns – external assistents such as the mathematician Jürgen Reichert from the MHD group were gratefully accepted – had already come to a standstill in 1982, the year after the final completion of the twin telescope. At the beginning of 1990 the Azerbaijan episode, which had fallen out of time, was officially ended by the ZIAP director K.-H. Schmidt.

Central Institute for Astrophysics Potsdam

Lauter's declining reputation as General Secretary and Vice President soon made his science-orientated three-pillar structure of astrophysics in the Academy obsolete, also because one of the directors, Wempe, was not member of SED. The Central Institute for Astrophysics Potsdam (ZIAP) was thus formed on 1 July 1969 already two years after the Institute for Stellar Physics was founded from the two observatories in Potsdam und Sonneberg. Now almost all astronomers of the GDR are assigned to one and the same director – the mathematician and science historian Hans-Jürgen Treder appointed by the Academy President. Such sort of hierarchy was the basic prerequisite of the SED for the patriarch-socialistic society so that indeed only one single person was responsible for all problems with (say) astronomers and administrators, even in cases of morale or political insubordination. Excellent science was never a condition for persons to be selected for such top positions, only loyalty and forcefulness. Even the Sternwarte of the University of Jena was envisaged as part of the ZIAP, which plan was cancelled, however, when Lambrecht in Jena was accused by his rector for violating the basic principles of the university management. He had "established contacts with relevant heads of other academic institutions with the aim of achieving changes with far-reaching consequences".[31] Lambrecht wanted to avoid the actual university reconstruction which would have made him a group leader instead of director. He resigned in protest and retired as a full professor of astronomy already in 1974.

Originally, the new institute's rules stipulated that *incoming foreign mail must be handed to the addressee unopened,* after which the addressee must inform the section leader of the content through official channels. *Outgoing mail must be signed for by the section director.* A short time later, this procedure was replaced by the much simpler soviet rule: *Postal traffic always via the director of ZIAP.* At a staff meeting on 13 May 1986, it was then stated that "letters to the FRG and West-Berlin are to be sent to the director in quadruplicate", an instruction allegedly due to the lack of photocopiers. All the major academy institutes formed comprehensive scientific combines that had also moved from the collegial leadership of an academic community to strict personal leadership at all levels of command. The five central institutes of the Cosmic Physics[32] Department of the AdW, were subordinated to a department leader and all department leaders to the President and his General Secretary. The classes and sections of the old AdW only played a decorative role in the future, the new elite liked to elect each other, and consequently there was soon the meaningful new name "Academy of Sciences of the GDR" describing the real situation because this organisation was entirely in the hands of the socialistic party leaders. In the regulations for the selection of the top cadres (*nomenclature rules*) of 1971, the first and foremost requirement for the personnel policy was to *secure the leading role of the working class and its Marxist-Leninist party*— not a single word about scientific expertise. Soon after the foundation of ZIAP Treder as an eloquent and versatile interpreter of the physics history gave a performance to all of his employees in Potsdam's largest cinema on "Why does the director of an astrophysics institute have to be a theorist?" after the text had been handed out in advance.

Also, his deputy Ruben had always seen himself as the head of a theory section. Until the formation of the Relativistic Astrophysics group at the Sternwarte Babelsberg in 1975, even the profiled relativists Dautcourt, Mücket and Müller belonged to his section. He had spent his school years in the German-Jewish academic diaspora in Ankara and later, interned, in Central Anatolia. In 1935, at Atatürk's request, his father had accepted a chair of Indology and lived with his family together with emigrants such as Ernst Reuter, Paul Hindemith and

---

[31] Archive Schielicke.

[32] Since 1974 Department Geo- and Cosmos Sciences chaired by Stiller 1973-1984, followed by Kautzleben.

Bruno Taut in the culturally closed anti-Nazi "Colony B", where the children, mocked by their Turkish peers as "haymatloz", were privately educated. After the war, his father travelled from Chile via Hamburg to East-Berlin, applied for membership of the SED in 1952 and started a stunning career: National Prize, institute director, academy member with own summer house on the privileged beach of Hiddensee. The son studied astronomy at the Humboldt University in Berlin, wrote his diploma thesis in Potsdam, and—possibly on Wempe's advice—soon took up an assistance with Alla Massevich from the Astro Soviet in Moscow where he obtained his doctorate in 1962 with numerical calculations on the structure of stars under the simplifying assumption of homogeneity. His Russian-language publications at that time have remained without measurable resonance, while Massevich's favourite Tutukov helped to determine the level of this field years later. As her interests lay more in geodesy and astrometry, she occasionally visited colleagues on the Telegraphenberg and probably also met Wempe on these occasions.

Johann Wempe wrote in the 1961 annual report of the AOP that the group leaders E. Lamla, W. Mattig and E. H. Schröter left Potsdam at the beginning of August and that U. Güntzel-Lingner did not return to Potsdam from the IAU meeting in Berkeley.[33] Güntzel-Lingner had conducted hidden negotiations with Walter Ulbricht in the spring of 1961 when, in a letter dated 10 June, he asked Ulbricht for permission to keep a private car that a very distant friend, who had in the mean time left the GDR, had parked in front of his house with keys inside and subsequently given to him in writing. He, Güntzel-Lingner, wanted to popularise Soviet space research successes even more intensively with this car. As there was no reply, Ulbricht had to bear the consequences and "GüLi" followed his very distant friend to West Germany.

Ruben completed his three-year doctorate in Moscow and was appointed to the position of senior assistant on 1 January 1962. Only a few years after getting the doctorate and with further stays in Moscow, Ruben became in Potsdam the first Managing Director of the new ZIAP. In 1973, he informed the Academy that the documents for his dissertations all remained in Moscow; he himself only had a simple receipt, but his diploma certificate was here in the Potsdam.[34]

For advent 1972, Rudolf Tschäpe who had just moved from Sonneberg to Potsdam after his military service as an unarmed soldier——together with Christians of all kinds, mostly Adventists, Baptists and Anthroposophist's— invited the members of the Theoretical astrophysics group to his new

---

[33] AOP staff 1960. Solar physics: Jäger, Künzel, Mattig, Schröter; Stellar physics: Günther, Güntzel-Lingner, Lamla, Oetken, Rolf, Schneller, Wempe; Solar radio astronomy: Böhme, Daene.

[34] Brandenburg State Main Archive (BLHA).

two-room flat in any of the decaying baroque blocks in Potsdam. Candles, cake and pastries from the bakery round the corner and advent music to the welcome. On entering the first room, one noticed a magnificent antique farmhouse board that Tschäpe bought in the Sonneberg region, and in the other room a golden-gleaming astronomical clock by Cuno Hoffmeister with moon phases and sidereal time, an early Wagner sculpture by Max Klinger and stacks of printed computer endless paper in all corners. Tschäpe had moved to Potsdam from Sonneberg because of the local computer capacities; it was exactly the beginning of the machine computing era in the applied physics branches. At that time, there were two powerful computers in Potsdam, a Polish Odra in a Shipbuilding Centre and the Russian supercomputer BESM-6 at the Meteorological Institute. More complex numerical dynamo codes were first developed on an NE-503 from National Elliott at the Dresden Institute for Data processing by Hiller and Recknagel. Hiller's ingenious code for solving a huge algebraic equation system was an early application of computing technology to astrophysical problems. After a failed escape attempt to West Germany, Hiller was sentenced to 8.5 years in prison for alleged espionage in an absurd trial; his name was never mentioned again in Potsdam.

Ruben looked through Tschäpe's bookshelves criticising the disorder, Tolstoy next to Christa Wolf, hm. He had known for a long time that Tschäpe was being investigated by the MfS in order to gather conclusive information on *anti-state agitation and group formation*. Ruben had to guarantee that no doctorate degree would be awarded to him. It had to be checked *whether Tschäpe did not have the opportunity to provide specialised information to West Germany*, his *role among the military service resisters* had to be worked out, demanded the MfS officer in the cover sheet for the operational personal file "Kontakt", *IM Astronom, supervisor of the OPK* was responsible for the success of these investigation.

Tschäpe had been observed in Sonneberg since he had been checked at the border crossing in August 1968 while travelling back from Prague on a motorbike with a friend. Both had enthusiastically told the investigator that there had been no political riot in the ČSSR, that the Soviet tank soldiers thought they were on a manoeuvre and were happy about the many flowers thrown at them. They themselves had been eyewitnesses and had brought newspapers with them as proof. The officer wished some things reported twice and the naive students recounted how everything had really been. After 13 hours they were released, everything in writing had been retained and a protocol text signed. After 25 years, Tschäpe's diary—in which English-language sentences had been translated into German in the easily identifiable handwriting of IM "Hagen" from the Sternwarte—will be found in the MfS files along with a precise floor plan of his two-room-apartment in Potsdam.

The year 1971 brought the Astrophysical Observatory a journalistic sensation. Paul Roberts and Michael Stix during a working visit to the HAO in Boulder, Colorado, had translated 14 German-language publications by the Krause/Rädler/Steenbeck group (now in alphabetical order!) and published them as an technical note, as they would otherwise be "hardly accessible" (Fig. 7.13). This implied that German-language publications were from now on considered as inaccessible, but it also meant that the AOP had re-emerged internationally. Not only Steenbeck had a high opinion of his work, but foreign experts also appreciated the new methods. Roberts held a chair at the University of Newcastle and Stix worked at the Universitätssternwarte in Göttingen. The turbulence-theory approaches explaining cosmic magnetic fields had not only impressed the two translators, it has remained in constant development worldwide over many decades. Since the friendly reminder of the two translators, all of the AOP's major spectroscopic and theoretical publications have been written in English.

From 1974 on, a turbulence-expert group promoted the GDR-wide networking of all scientists in this field, which led to regular one- to two-day colloquia in Berlin, Potsdam, Dresden and three more destinations. In 1978, a conference in Eisenach brought together turbulence experts from a wide range of applications, with contributions on channel flow, flows in stables and transport problems relating to the fluctuating motions in the Earth atmosphere and the sun. In contrast, contacts with the Central Institute for the Physics of the Earth on the same Telegraphenberg were never structured. Initially, there had been seminars with a small group from the former Geomagnetic Institute, of which F. Wiegank's impressive report on the connection between palaeomagnetic measurements and continental drift was particularly interesting. In the early 1970s, magnetic geodata made it increasingly likely that Africa and South America had separated from each other 150 million years ago. A late triumph for the continental drift theory of the former astronomer Alfred Wegener for which he once had to suffer massive and malice critics ("crustal drift disease") from professorial "experts". Wegener died on his 50th birthday during a retrieve mission to the ice centre station on Greenland. The young Wiegank had attended a lecture in Magdeburg in the early 1950s given by Wegener's widow about her husband's ideas, which were still ridiculed. Thanks to his awareness, we were one of the first groups in Germany (!) to learn about the latest findings on geomagnetism.

Instead of the Nobel Prize for Steenbeck there had been a 1st class National Prize in 1971, and Krause was promised by Treder the title of professor. The SED members of the Cosmic Physics Department had categorically demanded that Ruben should become a professor instead. M. Becker, first a baker then a

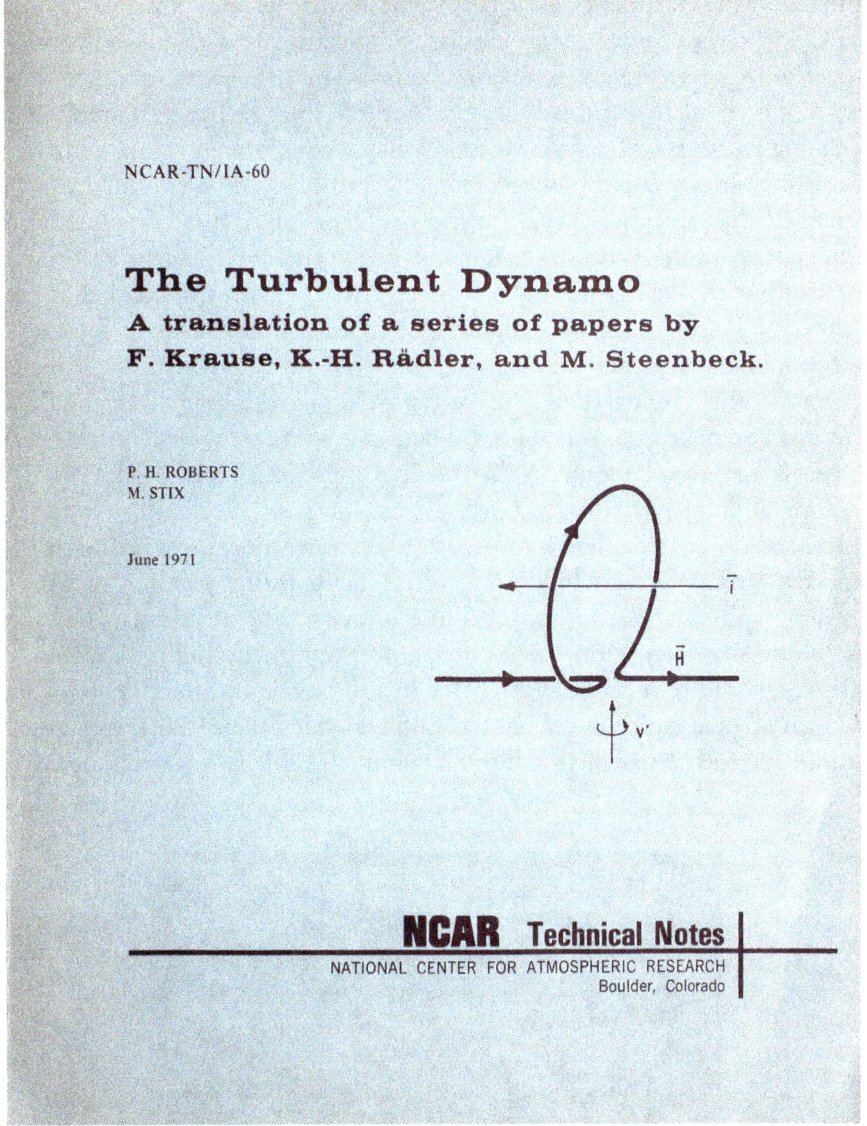

Fig. 7.13   The turbulent dynamo. A translation of papers by F. Krause, K.-H. Rädler and M. Steenbeck. Roberts & Stix (1971)

SED secretary, attested Krause a *negative attitude towards the politics of the party and state government*. The hefty and dreaded Pätzold as IME with the the self-chosen and much-promising code name "Kosmos": *I have recommended to the party group that they maintain their critical assessment of Dr. Krause, reject his professorship and support that of Gen. Ruben.* Pätzold was the security officer of the whole Department Geo- and Cosmos Sciences. The former SED functionary, who was completely unfamiliar with science, was to fulfil all of Stiller's tasks relating to security and control of astronomers, geologists and geophysicists. Like all such commissioners everywhere, he liked to get involved in personnel matters with and without Stiller's consent, he formed his own power centre and was highly feared. The intrigue, however, failed, Treder explained that Steenbeck would otherwise fight for Krause, so if he didn't, then neither would Ruben (Fig. 7.14). For his own safety, Treder had Krause's publications checked by one of his students. After years of disputes behind the scenes, the President finally signed the certificate of appointment for Ruben in September 1976 and a year later for Krause who was now sitting at Schwarzschild's original desk. The AOP, now known as Section II of the ZIAP, thus had two new professors, one with and one without own scientific results. "Dear Krause, dear colleague, it has finally happened...", Steenbeck started in his congratulatory letter suggestively.

In May 1975, Treder had informed at a very early morning meeting that he expected me to write a habilitation treatise the following year, in the form of a monograph on stellar rotation because of my youth. Before that, however, an "impressive" article on the 275th anniversary of the Berlin Academy of Sciences would have to be written for institute-wide information. This not too strong precondition was gladly fulfilled, but Stiller, the leader of the Cosmic Physics Department – himself coming from Jena—raged undercover.

Fig. 7.14 Hans-Jürgen Treder (1928–2006). (Photo M. Schulz-Fieguth. Courtesy of M. Schulz-Fieguth)

Rüdiger and Rädler will not habilitate, it was agreed of dismantling the working group of Dr Krause he dictated into the MfS protocols as IM "Martin".[35] Rädler had provoked the displeasure of Stiller because he had insisted that, as the chairman of the institute's personal committee (since 1974), he would only deal with social issues rather than with political ones.

Stiller's intentions finally became a program. In fact, in his Concept for the ZIAP at the end of 1981 the dynamo group together with stellar spectroscopy only appeared casually. Due to the spin-off of the AOP instrument-construction group from the ZIAP, the spectroscopists remaining in the refractor building had lost their technological roots and were later united with the Einstein Tower team and the Tremsdorf radio observatory under management of somebody who was willing to construct the detailed office plan. The photometry group will also leave Krause's "Section II" after he had not supported their overdrawn claims to academic honours; the original scientific structure designed by Wempe has finally been reduced to the tiny dynamo group. A list of the ZIAP's alleged top results compiled by the director reflected the hierarchy of power. All medal ranks were occupied by Treder's works, including the "quantisation of gravitation fields", which he claimed as indispensable for astrophysics. "Dynamo theory" and "Structure of the [solar and stellar] magnetic fields" of the former AOP follow in seventh and eighth place. The external evaluations carried out in 1990/1991 will not follow this ranking but will actually turn it on the head. Treder's later workplace in Caputh was one of the few institutions of the AdW that did not survive the evaluation by the German Wissenschaftsrat.

In 1985 Treder delivered a publication of a theory entitled Aberration Constant and Rotation of the Cosmos. He determined the aberration angle during the annual orbit of the Earth for a distant object moving parallel to the earth with a different speed as supposed to be the result of a global rotation of the cosmos. A counterargument had just been published in a NATURE article that any primordial rotation of the universe would have been lost due to inflationary expansion. Nevertheless, Ruben immediately formed an astrometry group to provide empirical evidence of the universe's global rotation. The group had originally been founded in order to use the Tautenburg 2-m-mirror to measure the possible proper motions of quasars. It published a programme for connecting the data from the future HIPPARCOS astrometry satellite to a quasi-inertial system defined with distant galaxies in order to interpret the absolute proper motions of the stars. However, the astronomical satellite was only launched late in 1989 without the participation of German astronomers,

---

[35] Appendix 7.

and it has also remained unclear what the rotation of the HIPPARCOS system should have to do with the rotation of the universe. Finally, the application of the aberration formula to observations did not produce any results.[36] On the other hand, high-precision position measurements connecting compact galaxies could be used to determine the proper motion of objects on Tautenburg photo plates from different epochs to distinguish extragalactic objects from galactic ones. After 18 years, 10 of 36 blue and faint objects were found with measurable proper motions on Tautenburg images which are therefore certainly not quasars but white dwarfs.[37]

It was also according to Stiller's concept that in 1984 half of the gravitation staff were removed from the ZIAP to the Einstein summer house in Caputh as a new Einstein Laboratory for Theoretical Physics. As he had promised to the MfS, Krause's network was destroyed, the dynamo group was outsourced to a closed border frontier zone and Treder was dismissed completely. After such successes as a powerful scientific manager, Stiller's rise to Vice President of the AdW was unstoppable.[38]

Time Signals

In keeping with Jena's student traditions, the Astrophysical Observatory had presented from time-to-time cultural evening events, mainly with public readings by prominent, sometimes colourful authors, starting with the disgraced Christa Wolf. The visit on the Telegraphenberg was her first invitation since years, which made us naive organisers proud rather than thoughtful. I had already met her in 1969 at a literature seminar in a Potsdam city centre. At the Telegraphenberg, she had read from a draft of "Kindheitsmuster", which later became a famous GDR-wide symbol for being with friends due to its eye-catching cover on bookshelves. "The past is not dead; it is not even past", the provocative text begins. The actually reserved Wempe took great care of the Wolf's, made his official car available to pick them up and drive them home, and then took the opportunity to entertain Mrs and Mr Wolf in his office with its ceiling-high old wooden cupboards. Stiller then dictated in MfS files: *Due to Johann Wempe's negative attitude, the entire Astrophysical Institute can be categorised as a negative group.*[39] In fact, even Wempe suffered

---

[36] Ruben (1991).
[37] Richter & Scholz (1990).
[38] 1984–1988.
[39] Buthmann (2020), p. 1116.

from the system. Before August 1961, on 1 June, he reported to Cuno Hoffmeister, chairman of the astronomy section of the AdW, that he had been denied permission to attend the colloquium on astronomical observation techniques in Tübingen, which began the next day, on flimsy grounds. *As I do not intend to attend any other conference this year, I considered a rejection of this modest request to be quite impossible,* he wrote.

The visits of the famous economist Jürgen Kuczynski, the GDR star architect of Stalin boulevard and Jena Tower Henselmann, the high-talented author Günter de Bruyn, the ironic writer Renate Holland-Moritz and the young DEFA film directors Rainer Simon and Lothar Warneke were the most memorable highlights within the series called "Time Signals". High impressive was also our encounter with the later honorary professor Carlfriedrich Claus in his worn-out flat under a cinema somewhere in Saxonia, who could easily pass as an avantgarde eccentric. At night in trance, he produced his unique "language sheets" on parchment, which served as a signal of artistic-oppositional independence to their numerous buyers. His work would later be declared a registered cultural asset of state Saxonia. Or the visit to the young painter Uwe Pfeifer in Halle-Neustadt, who had placed a grey huge building in a grey fog, a few windows lit up, rigid anonymous men with hats standing on the paths, the street lights switched off, only silence to be seen. There was a special cultural fund at the institute, the money from which was used to buy this artwork. An aggressive double self-portrait by Volker Stelzmann, which we also acquired had to be returned to Leipzig after a negative vote by the Observatory staff. Harald Metzkes in Berlin, Arno Rink in Leipzig, Theodor Rosenhauer in Dresden and Thomas Ranft in Chemnitz also had ample opportunity to marvel at the art-loving young astronomers who drove up in a large official car from the AdW. However, the revival of a free cultural scene—naive, public, independent and without any financial requirement – came to a standstill after the expulsion from GDR of the famous singer Wolf Biermann in November 1976.

The early installation project "The Light" by the young Leipzig artists Dammbeck and Grimmling for the midsummer night in the dome of the Great Refractor failed due to the mistrust of the institute's directorship (Fig. 7.15). Participants were supposed to wait together from midnight to sunrise in complete darkness for the first morning light, which would only hesitantly reach the exhausted people and their rusty giant telescope through the gap in the dome, which opened very slowly in the right direction at the right time. The security chief Pätzold of the Geo- and Cosmos Sciences Department ordered at the end of 1982 that "exhibitions on the Telegraphenberg cannot be made accessible to the public anymore". This

Fig. 7.15 Closed observatories; left to right: P. Rohn, R. Tschäpe, G. Tschäpe, G. Rüdiger, S. Ketzscher, L. Dammbeck, H.-H. Grimmling, Children: Karl-Konrad and Elsa Tschäpe, ca. 1976. (Photo K. Plessing. Courtesy of K. Plessing)

order ended all connections between the astronomers on the Telegraphenberg and the small, lively and creatively rebellious part of the GDR art community that Tschäpe and his friends had always been concerned with.

This ban came late, however, perhaps because parts of the SED district leadership were impressed by the wide and positive resonance of the Wieland-Förster exhibition in the dome of the Great Refractor. It was the first major exhibition of artworks of the 44-year-old sculptor Wieland Förster which took place on the Telegraphenberg in early summer 1974 (Fig. 7.16). The opening speaker mentioned the young scientists' realisation that their "research cannot only take place in the factual space of digits, but requires human imagination and social control as manifested in art". The now legendary event in Potsdam indeed marked the exact turning point towards the dominance of the computer over philosophy and society.[40]

At the request of the organisers and the artist, a spectacular sculpture was to remain on the Telegraphenberg; it could easily have stood next to the refractor building and the Einstein Tower. However, Treder, who remained completely alien to any modernism, did not accept that. Never again had the

---

[40] Dick & Ackermann (2025), p.421.

Fig. 7.16 Wieland-Förster-exhibition 1974 Telegraphenberg Potsdam; left to right: Tall Striding Man (1970, bronze, 190 cm). Tschäpe (1943–2002) at the opening speech. (Photo H. Pölkow. Archive W. Förster. Courtesy of W. Förster)

Academy of Arts and the Academy of Sciences come so close as at this exhibition. Both sorts of intellectuals were to remain separate and not develop common subversive positions. The local MfS reported to the headquarter in Berlin that *for the first time in our district, a section of the Academy of Arts is officially presenting itself in a section of the Academy of Sciences.* And anyway, *Initiator Tschäpe is being investigated for anti-state agitation, decline of military service and currently undefined connections to evangelic circles.* The MfS found both the artist and his presenter *as representatives of the opposition to the real existing socialism in the GDR,* it categorized everybody after their own political coordinate system. Later, for his commitment to the transition to democracy in East Germany Tschäpe was awarded the Cross of Merit of Germany in 1995. He died 2002 after a long illness shortly before the completion of his new home near Potsdam.

In November 1984, Grimmling, Dammbeck and friends sent out a clear signal against the official cultural scene with their semilegal "Herbst Salon" in Leipzig. A large floor in an exhibition hall, rented under a legend for a high price, formed the epicentre of independent art in the GDR for three weeks. The exhibition was closed on 7 December; an official vacation order had been postponed several times for fear of causing a land wide protest. Shortly later the autonomous cultural scene everywhere was forced to retreat under the roofs of churches. In early 1985, the year of their westward escape, Grimmling managed to erect a huge paper installation in the dome of the Nikolaikirche in Potsdam which was mercilessly removed after a few days. No wonder that the artist later found himself in a long pessimistic black period as a result of his experiences in the GDR (Fig. 7.17).

Even earlier, the state marked the organisers of the Time Signals from the AOP as enemies and planned to punish them with lifelong professional isolation. It didn't last, however, that long. Already on 7 October 1989, the 40th birthday of the GDR, there were unannounced demonstrations in cities such as East-Berlin, Leipzig, Plauen, Dresden, Jena and Potsdam, which were broken up with about thousand arrests. On 9 October, more than 100,000 demonstrators calling "Wir sind das Volk"[41] will set off unchallenged in the

Fig. 7.17 "Together". Painting by Hans-Hendrik Grimmling 2001. (Courtesy of H.-H. Grimmling)

---

[41] "We are the people".

dimmed city of Leipzig. As a result—and not through clever activities of the Academy representatives as it had been their duty—the two-class society of AdW that has carefully been constructed over two decades by SED science commissioners will implode for ever.

Everyday Theory Except Sunday[42]

"*By the end of 1972*" it is reported, "*Dr Krause had already been to Great Britain three times and Prof. Roberts had already been to Potsdam three times.*" At this time, there was a veritable dispute between Krause/Roberts and I. Lerche from Chicago with H. Abt as the editor of the Astrophysical Journal. Krause wrote in a 1973 report that Lerche "makes sensational claims which he proves with incorrect calculations". The two friends developed rigorous proofs for their point of view, mostly after applying the old Bochner theorem according to which spectral functions of stochastic processes are positive-semidefinite. Krause, who could calculate any solvable integral on the blackboard without a noticeable pause, mainly worked with people from England, perhaps because their research profile of applied mathematics came closest to his preferences. On 18 April 1969, P. H. Roberts wrote to the director of the ZIPE that Krause's presence at a recent conference in Newcastle would have been of the greatest value and suggested a valuta-neutral exchange between the groups in Potsdam and Newcastle for the future.[43] The letter was probably given to Krause on his return way from the meeting home. Since the academy reform, permanent relations with Western countries were no longer part of the job description of employed scientists. The few confirmed travel cadres occasionally attended international scientific events, but were rarely recognised there. One of the few GDR delegation leaders complained about his colleagues' poor English, which *made it impossible to contribute anything to the conference.*

Also, the fact that things went differently with Krause including organising his own international meetings with prominent participants on the Telegraphenberg[44] and later in the Sternwarte Babelsberg[45] brought him dangerously close to the security forces. They were by no means only interested in foreign affairs, as he perhaps had initially assumed, but almost always mainly for his immediate private and professional surroundings, a dilemma from

---

[42] Advertisement of a car driving school in downtown Potsdam. M. Knölker found this phrase typical for all dynamo groups worldwide.

[43] Appendix 6.

[44] Conference 1983, "Stellar and Planetary Magnetic Fields", AOP Telegraphenberg.

[45] Conference 1988, "Magnetic Fields in Galaxies", Sternwarte Babelsberg.

which he could never escape. The secret talks took place in the mornings in his private home when his wife was absent. He was regarded as *difficult*, preferring to explain to the officers the latest successes of his co-workers, explaining the quality of them and their scientific position on a global scale. In May 1976, a dispute arose over *his reporting style*. It concerned his fears that *reporting on political statements made by individuals would only harm them.* Krause had avoided private contacts with his colleagues as much as possible; not knowing anything, not visiting anyone and keeping his surroundings at a distance were his preferred strategies against the secret officers. IM "Schmall" only reported what he had heard or seen by chance. In 1980, IM "Jochen Gränz" complained in a written assessment that "Schmall" did not take the initiative to solve the tasks assigned to him. Once, at the end of a written meeting minute, one agent wrote resignedly: *In the* **XXXX** *case, the IM is clearly speaking the untruth.* The final report on the termination of his IM activity due to illness/age is from June 1989, the *IM was not willing to penetrate the private sphere of persons on behalf of the MfS.* In another case around 1985, the astronomer IM "Dr Mann" trained officers of the MfS on how to evade Western security checks as a fake astronomer with real Tautenburg photo plates under his arm and researched where passports were kept overnight in hotels. It is unknown whether in the following years such a hidden operation took place after this scenario.

"Explain the differential rotation of the sun, find the α effect of hydrodynamics," Krause had told me in 1969 as his first and only doctoral student. The solar equator rotates faster and overtakes the polar regions every 100 days. The search for this so far unexplained phenomenon soon revealed as a consequence of rotating turbulence, through which rotational momentum constantly flows equatorward leading to a higher rotation frequency there. Large turbulent regions of stars can thus never rotate like a solid body. Such turbulences overcome the law of experimental physics, according to which angular momentum only ever flows into areas of slow rotation, so that in the end the star would have to rotate rigidly. The newly discovered effect, which had remained unknown in terrestrial laboratories due to the scales, was first published in a series of articles in the journal Geophysical and Astrophysical Fluid Dynamics founded by Roberts who edited GAFD from 1976 to 1991, succeeded by A. Soward. I called this turbulence phenomenon the "Lambda effect", in a hidden allusion to Ludwig Biermann who had already drawn attention to the differences in the angular momentum transport of isotropic and anisotropic turbulence in 1951.[46] At the end, the faster rotation of the

---

[46] Biermann (1951a, b).

solar equator – seen already by Carrington and Spörer—results from nonlinear interactions that cannot be understood with linear physics.

In addition to P. H. Roberts, A. Soward, K. H. Moffatt, N. Weiss, O. Lielausis, A. Gailitis, I. Tuominen, L. L. Kitchatinov and A. Ruzmaikin[47] visited the dynamo group in Potsdam, also in continuation of the traditions at the old Observatory[48] and the Steenbeck-Institute in Jena.[49] All talks related to dynamo theory, which was still young at the time; often the calculus were analytical without the aid of pocket calculators or even computers. They usually took place in Krause's study——once Karl Schwarzschild's office—in front of two newly installed huge blackboards, the lecturers with chalk in one hand and a sponge in the other.[50] During his visit to the AOP in 1976, Keith Moffatt worked on the proofs of his book.[51] He even visited us—illegally— privately at home, presented our young son with a still unforgotten red toy bus, and visited with us the nearby Cecilienhof Palace, the site of the Potsdam Conference of 1945. I had proudly told him about my latest result, that in rotating fluids the total turbulence-induced electromotive force for strong magnetic fields disappears inversely quadratically, which, as I only realised later, he had already discovered two years earlier.[52]

Ivan Cupal, originally as guest of a neighbouring institution on the Telegraphenberg, brought together the dynamo people he knew from East and West in his Czech home town Alsoviće in 1979, including R. Hyde, S. I. Vainshtein and S. I. Braginsky who, despite or because of his prominence, was allowed to cross the Soviet border for the first time on this occasion (Fig. 7.18). In addition to the actual geophysical dedication of the meeting (Dynamo theory and the generation of the Earth magnetic field), the Potsdam group also presented latest results on the suppression of nonaxisymmetric magnetic fields by differential rotation (Rädler) and the shape of the α tensor for very fast rotating stars (Rüdiger). For the first time, at the age of 35, I was allowed to present my own results to an international audience. In the previous year at the well-attended workshop on Solar Rotation in Catania/ Sicily—organised by G. Belvedere and L. Paternò – it had to be done for me by the participant D. Gough in his distinctive way as I was told later. In his

---

[47] Later: Ya. B. Seldovich, S. I. Vainshtein, H. Yoshimura, D. D. Sokoloff, A. Brandenburg, D. Moss.

[48] Unsöld, Kuiper, Milne, Waldmeier, Kiepenheuer, Korsching, Strohmeier, Lyons, Krug, Houtgast, Chandrasekhar. H. Brück reports seminar lectures by Bok, Oort, Elis, Strömgren but did never see Einstein on the Telegraphenberg (Brück 2000).

[49] Arzimovich, Kirko, Stenflo, Velikhov, Keldysh, Mestel (incomplete).

[50] See Appendix 10.

[51] Moffatt (1978).

[52] Moffatt (1973).

Fig. 7.18 Alsoviće 1979; left to right: Bucha, Ševčík, Cupal, Boda, Gailitis, Brestenský, Hyde, Braginsky, Roberts, Trešl, Soward, Rädler, Ruzmaikin, Rüdiger, Hejda, Moroz, Ivanova, Vainshtein, Krause, Prochazkova, Horáček, Bräuer

closing remarks in Alsoviće, Krause was moved to formulate that "colleagues who had never seen each other before met here and now know how they think, which can never be achieved by reading publications". Also, in the accounts of the lifework of Paul Roberts the Alsoviće meeting plays a prominent role[53]; several authors of his periodicals from East and West were able to meet him there in person.

Before the trip to Bohemia, I had to prove my suitability for business trips to Eastern Europe with a visit in May 1978 to the Institute of Physics at the Latvian Academy of Sciences. It was Steenbeck's technological base camp for sodium experiments. The institute in Salaspils seemed to be a closed facility as visitors were brought there daily from Riga in private cars. It was the beginning of the midsummer nights in the north and Riga proved to be an amazing open-air architecture museum. The occasional visits to popular city restaurants in the evening led past long lines of waiting people straight to the entrance——they were Russians, according to our Latvian hosts. Although the Russians were the dominant political force in the Baltic states, they waited patiently to be let into the dining paradises long time after the Latvians.

According to Steenbeck's suggestion, a helically twisted channel had once been constructed along an external magnetic field in the MHD-laboratory at

---

[53] Soward (2023).

Salaspils,[54] through which liquid sodium was forced at high speed in order to measure the resulting electrical voltage along the field.[55] I gave my first English talk there on the eddy viscosity tensor subject to large-scale magnetic fields by the then newly appeared hand-written transparencies of poor quality. By a new MHD experiment, a huge rusty steel container full of mercury which was made to move rapidly with a motor-driven internal stirrer, the temporal decay of a flow vortex was to be studied. It was observed that strong magnetic fields extend the life of the vortices and a theoretical explanation via the magnetic suppression of the eddy viscosity was sought. Later, the Riga dynamo experiment with liquid sodium emerged from this monster but the turbulent effects always remained small (Fig. 7.19). It was only when inner baffles were installed to redirect the flow in the sense of a meridional circulation that an exponential increase in the field was achieved at the highest rotation frequency on 11 November 1999,[56] at the same days as the first magnetic field excitation in the new dynamo machine at the Karlsruhe Research Centre. As required for non-turbulent flows, the poles of the magnetic fields induced in such dynamo machines were always perpendicular to the symmetry axis despite the essential differences between the two experiments. Also, for rapidly and rigidly rotating turbulent stars, the calculations had shown the self-excitation of a magnetic field with poles perpendicular to the rotation axis.[57] The spectroscopists in Potsdam had consequently begun to replace the prevailing model of the "oblique rotator" for magnetic stars with combinations of simple basic solutions of the dynamo equations. The application of this concept to the well-observed star α^2 CVn by Oetken was immediately successful,[58] as it was with the 16 other magnetic stars available at the time (Fig. 7.20). In addition, Alexander Hempelmann had demonstrated later that a large dark spot on the equator of the star HD 24712 could actually produce the observed light curve, too.[59]

At this time, the peculiar A-type star γ Equ, later famous for its short-living oscillations, was also under discussion whose magnetic field strength developed slowly with a polarity changing in 1970/71 and with a fixed radial velocity. The only remaining explanation was an atypically slow rotation of 72 years or an alternating field dynamo similar to that of the sun. Spectroscopic

---

[54] "I visited the Riga MHD laboratory in Salaspils, at that time the best of its kind in the world." Moffatt (1982).

[55] Steenbeck et al. (1968).

[56] Gailitis et al. (2000).

[57] Rüdiger (1980).

[58] Confirmed later by Gerth et al. (1999).

[59] in: Cowley et al. (1986).

Fig. 7.19 Dynamo experiment with sodium at the Institute of Physics Salaspils (2005). 1.5 m³ sodium, propeller-driven flow up to 20 m/s, motor power 200 kW. (Photo Th. Gundrum. Courtesy of Th. Gundrum)

measurements, however, had led to a rotation period of less than one month. For the supergiant v Cep, the rotation period would only be 5 years if the observed temporal variations of the strong magnetic field of 2500 Gauss[60] were of fossil origin or could be attributed to a stationary or slowly drifting nonaxisymmetric dynamo. The latter would only be possible for fast, almost rigid stellar rotation. Perhaps because a unified picture of the data was then

---

[60] Scholz & Gerth (1980).

Fig. 7.20   Left to right: K.-H. Rädler; L. Oetken; F. Krause. (Courtesy of M. Krause)

not achieved a summarising presentation of the observations in continuation of the book by Krause and Rädler about dynamo theory[61] was never written.

Around 1985, the well-known magnetic star β CrB attracted widespread attention. Oetken had noticed that this star had left the Main Sequence so that energy production was already taking place in the stellar envelope.[62] Is this possibly the case for all known magnetic stars? Another open question for the still tiny dynamo community, many of which were posed in 1985 at the IAU colloquium in Simferopol, Crimea, the first meeting on magnetic stars that did not suffer from political restrictions (Fig. 7.21).

Polit Astronomy

In the 1970s, a doctorate at the Academy included a written discussion of a scientific-philosophical problem. On the instruction of Ruben, I was not allowed to ask the respected Jena philosopher Lesser for a report although my

---

[61] Krause & Rädler (1980).

[62] in: Cowley et al. (1986).

Fig. 7.21 Simferopol 1985, all names given by Cowley et al. (1986); Potsdam participants left to right: Schöneich, Musielok, Hempelmann, Krause. Oetken and Gerth are invisible

study about the direction of time in theoretical physics referred to earlier disputations at his institute. Rather, the assessment was in hands of a Mrs Goetz from a college of education in Potsdam, who's full-time job of the leader of the GDR-socialistic youth organisation of the institution the current professor had previously been. My study about the preferred arrow of time— although the classical equations of physics do not imply a difference between past and future—was rejected because Lenin did not appear! A second version, later produced after reviewing a cheap textbook ("Lenin and science") with arbitrarily placed political quotations, finally was mildly accepted.

The organisers of the mandatory one-week Marxist-Leninist training course for external candidates at the same institution, during which frank discussions was expressly encouraged, were more pragmatic. After two or three of my oral contributions, which were probably still characterised by the Jena student atmosphere, a delegation of older teachers – who hoped to escape their profession by doing a doctorate – suggested that they would send me the necessary certificate if I left the course in the middle of the week without discussion. I announced only to leave after reception of the paper. On Wednesday, I was indeed back home at lunchtime with the stamped and signed certificate. It wasn't the culture of discussion I was familiar from the Jena university, but the attitude of the course managers was much closer to "live and let live" than that of the former FDJ secretary, who had strictly insisted on several exotic Lenin-references. The woman must have been aware that the situation was not without danger for me: without her signature, I would not have been able to complete the doctorate and at best I could move to Sonneberg if the institute's leaders in Potsdam had agreed. A doctorate position was a temporary qualification programme, not an employment contract. There was only one way out: Lenin had to be included in the theoretical text without noticeably lapsing into parody. I had fallen over in contrast to Erwin Freundlich at his time. Sonneberg instead of Istanbul, the big world had become tiny for young East German astronomers.

It was precisely these poles between which we had to live, something that the engaged Horst Lesser in Jena was not given the opportunity to move between the two poles and who, in these days, fell silent forever. In 1966 his boss at the Philosophy faculty had helped our Physikerball, which was always organized by the third year students, become legendary. During the performance of the singer Wolfgang Dehler from the Weimar National Theatre, who moderated and sang slight-politically suggestive songs about red and black colours, dressed in the same way, the philosophy professor sat backwards on his chair in protest, which became the talk of the city the next day. Dehler had once picked me up in his car on a country road, where he demonstrated his knowledge of all the East German dialects, and immediately agreed to appear at the famous Physikerball at the Jena university. When he later heard about the political difficulties of the organisers, he said that it was a must for him to perform in Jena for his reputation, and he would have gladly done so.

The Physikerball in May 1966 was one of the highlights of the academic year at the university. In preparation, half the city was covered in Dadaist slogans, a film was shot, the political number programme was written and well-prepared. Some of the students were surprised to realise that they could have been also theatre intendants instead of laser physicists. In retrospect, there were serious political difficulties for the student programme group, but these eventually went unpunished, partly due to the reputation of the students involved and under the protection of individuals such as Dr. Lesser. Maybe it was because of this positive experience that I only understood my own country when it was too late, at least not yet during the first Potsdam years. Much later, the files from archives revealed that there had even been an expedition of Potsdam MfS agents to Jena to determine my (rather minor) personal role in the preparation of the political cabaret and to find out whether some physics students had once conspired to set up together on Potsdam's Telegraphenberg for subversive purposes.

On 5 August 1968, the MfS reported to Honecker that an employee of the Geomagnetic Institute on Potsdam's Telegraphenberg had used a holiday trip to Bulgaria with a friend from the Sternwarte Babelsberg to escape from the GDR. When questioned, director Fanselau replied that the refugee had joint a room with the INTERKOSMOS expert Möhlmann, but without insight into his secret collaboration with the Soviet Union. Fanselau emphasised that he had planned to fire the arrogant doctor after his vacation. Treder's reaction is reported in a similar tone: Harald Fritzsch, talented but full of himself and prone to overestimating himself, had no insight into the important scientific problems of the institute. He was in possession of his diploma thesis, which he probably had with him when he fled, Treder said. The thesis dealt with problems, which he, Treder, regarded as a purely formal theoretical topic. What the Potsdam MfS division was apparently not yet aware of was the leading role played by the two refugees in the protest against the demolition of the Leipzig University Church on 30 May 1968 ordered by Ulbricht. At the final concert of the International Bach Competition, when the last speaker had finished, a huge yellow flag with the inscription "We demand reconstruction", a black cross with the symbol †1968 and the silhouette of the church was automatically unfurled. This was a dramatic public event and the security officers interrogated dozens of music, theology and graphics students in search of the actors—they, however, never thought to look for young physicists in Potsdam. Rudolf Treumann from the same institute for geomagnetism, who had painted the giant flag, followed with his family to West Germany only in 1978. There it became known that the hitherto anonymous fifth conspirator was none other than the discarded author Peter Huchel from Potsdam-Wilhelmshorst. The security officers in Potsdam had gained a first impression of the explosive power that culture and science can generate together, and they would experience it still several times in future.

On 23 January 1975, Ruben gave the Sonneberg astronomer Isolde Meinunger a written warning,[63] because she had attended Symposium No. 67 of the International Astronomical Union on variable stars in Moscow in July 1974 during her official vacation, with a tourist visa and at her own expense but without the permission of the institute administration. Because the obser-

---

[63] Gürtler & Dorschner (1992).

vation of variable stars was her field of work and her employment contract required her to undergo further training, Meinunger brought a protest against the reprimand at the Potsdam administratic court. In a written statement requested by the court, Ruben wrote that "she should realise that the diverse international relations of the AdW of the GDR and its activities within international organisations and associations are part of the tasks of the central administration. The most aspects of international scientific work—not least security interests—play a role here." Because of the security interests, the judge knew how to decide though Ruben would not have been able to name a single security risk resulting from Isolde's appearance at the conference of astronomers. Not even the possession of a passport was enough to practise the astronomer profession abroad, no, it was the permission of the authorities that was needed in any case. Only in 1989, in the Humboldt film by Rainer Simon[64] the young Humboldt shouts bluntly why he has to ask a king where he may travel, with his own money! The censors had already resigned.

After this sentence, even attending a public Sunday lecture at the Leopoldina Academy in Halle would have been subject to authorisation for AdW employees for reasons of security—so quickly had the Steenbeck-doctrine of 1967 led to a totally controlled science organisation. Meinunger's professional confession was 100% correct and what would have become of the AdW nomenklatura system if the judge had decided in favour of her? In June 1983, IM "Hagen" informed the MfS that in Sonneberg both Meinungers as well as Bräuer and Rößiger had refused to sign an official declaration of protest directed against *President Reagen's anti-human plans to misuse the outer space enviroment for weapons of mass destruction.* The informant's legendary refusal to work on astronomical observations was apparently compensated for by the intensive transmission of private observations. Still in May 1989, I. Meinunger was labelled an obstinate *cabin voter* by the hidden control group because she had used the only cabin in the polling station that had been set up by the election commission. Everything went wrong in East Germany.

A conference with a wide scope was scheduled for August 1975 in Prague: Symposium No. 71 of the IAU on Basic mechanisms of Solar Activity.[65] Originally, the initiator Bumba only wanted to deal with observations of solar activity but then preparatory discussions, primarily with Kiepenheuer from Freiburg, led to the opening up of the agenda, also because of the new Potsdam turbulence concepts on the rotation and the magnetic fields of the sun. I had

---

[64] The Ascent of the Chimborazos (DEFA/ZDF).
[65] Bumba & Kleczek (1976).

registered a talk on angular momentum transport by rotating turbulence and immediately received an invitation from Chairman Parker to open the session with my presentation. There were no difficulties to be feared as the publications had appeared or were in print, the accommodation had been ordered, the slides well-prepared and the English text of the talk memorised. Director Lauter from the ZISTP had nominated seven participants from Einstein Tower and Radio Observatory Tremsdorf, almost a school trip—what else could happen on the part of the ZIAP? At the time, the ČSSR was just the most socialist country in the Eastern Bloc, so why should Bumba be annoyed? However, in July 1975 the actual SED secretary of the ZIAP met me by chance (or not?) in the corridor of the main building of the observatory and demanded that I must personally cancel my lecture and hotel room because I had been removed from the list of the conference participants. Instead Dr Oetken would attend the meeting, which she did indeed without comment, she would then be absent from the next IAU colloquium Physics of Ap Stars in Vienna. As a protest, I had refused the demanded personal cancellation of my participation. *Rüdiger and 'Schmall' don't know the reasons why Rüdiger was removed from the list of participants at the symposium in Prague. He still suspects that other scientists are responsible for that*, the MfS triumphed. There was indeed no noticeable reaction from colleagues to this discrimination, only the head of the Einstein Tower, Jäger, briefly had taken me aside in the library to encourage me. Thanks to the restless efforts of the secret agents, the international premiere of my personal contribution to understand stellar rotation laws did not take place in front of an audience of hundreds in the Czech capital Prague in 1975, but only four years later in the tiny town of Alsoviće for just 20 people—probably not the only operational success of the invisible gentlemen.

At the end of 1975, a Major Wolf summarised that *the group of people being investigated was an oppositional group with anti-state tendencies whose activities remained below the level of criminal relevance. This group included the persons Tschäpe, Dr Rüdiger, Freigang and* **XXXX**.[66] *... Due to inadequate personal policies, the situation was allowed to develop because graduated students were employed as scientists who are not members of the SED.* The operational files were archived by the MfS after years of monitoring the naive organisers of innocent cultural events.[67] Although the group was not arrested for its activities, it was only tolerated as an *opposition group* at the AdW in the future. The officer had generally considered the existence of non-SED scientists in academy institutes as

---

[66] Name darkened by the BStU.
[67] MfS BV Potsdam Dept. XVIII, Reg. no. iV/1003/72.

a political mistake. The ideology of the reigning class could not have been expressed more clearly. The suspicion that party membership among academics in the future GDR could become the rule—just as it had been for the North Korea delegation—began to intensify with the recruitments during the 1980s. Since then, the granting with an additional (small) academy pension, to which all scientists had previously been entitled after only two years as an assistant, was only accepted in exceptional cases following a special application by the institute directors.

The proceedings of IAU Symposium No. 71 contain an informative summary of the general discussions by Durney, Gilman and Stix with the overall conclusion that "theories of differential rotation and dynamo theories are closely linked". This was precisely our concept, but there was also room for arguments against the idea that "mean fields" may exist in stars at all. At this conference first results of O. C. Wilson of magnetically induced stellar calcium emission for 7 stars were also presented, measured over almost 10 years on Mt Wilson. Three of them showed solar-like cyclic behaviour with periods of about 5 years, while no fluctuation was seen in three others. This perfectly matched the recent rediscovery of Spörers grand minimum by J. Eddy who had dispelled the last doubts that sunspots had disappeared for several decades after their discovery in the 17th century.[68] Since the meeting in Prague, more and more data from stellar observations have flowed into the solution of the solar magnetic field puzzle and the construction of turbulence models. Schwarzschild's vision will be fulfilled; solar physics becomes stellar physics—but not yet in Potsdam because this message had not really reached the Potsdam participants. After a few years, the Astrophysical Observatory on Telegraphenberg will even be closed down completely and most researchers will move to the Sternwarte Babelsberg. Treder who had once also unsuccessfully resisted moving to Babelsberg, said laconically that *one* mountain is easier to defend than *two*—failing to recognise his own situation.

In the summer of 1989, the General Secretary of the AdW will begin a letter to the directors of all central institutes with Dear Genossen[69] and think nothing of it; even a female Genosse did not exist there, and certainly no neutral director since K.-H. Schmidt of ZIAP had joined the SED in 1986. Shortly before, the stressed functionaries had come up with the idea that having permanent visas for travelling to the West would be more appropriate for their hard work than the stressing task of obtaining new stamps each time. Too late, soon afterwards every GDR citizen had unlimited travel authorisa-

---

[68] Eddy (1975).

[69] SED members addressed each other as "Genosse".

tion and the Academy of Sciences had an existential problem, the only solution to which—radical reforms—was not found in time by the authorities. In 1991 almost all of the former professors disappeared from the scene, mostly for reasons of their missing scientific qualification—everywhere the middle management of the Academy took the responsibility for the survival of the institutes.

8

Decays and Liquidations

The Solidarność Turnaround

During the day, English texts were read about violence, poverty and lack of prospects in today's Great Britain and at night, the sound of tank tracks recalled childhood days in the post-war hometown Dresden. The two-week language course for the current candidates for postdoctoral qualification of the AdW had begun on 1 December 1980 near the Oder-Neisse "peace border". In late summer, on the other side of the Oder river, the Polish independent Solidarność labour organisation had succeeded in legalising with strike actions under steering of Lech Wałęsa. The frightened socialist top leaders decided their resistance in Moscow on 5 December, putting an army tank group in the western bank of the Oder river the following day in order to create war noise on both sides of the border. The large majority of the participants in the course had formed an SED party group with a secretary for twelve days after the first evening meal in the dining hall. The few neutral participants had to leave the room. Krause told of a seminar for leading managers of the AdW where he was the only one to get up leaving the room on Monday evening.

Under the sound of the tank chains, the evening red wine groups that had soon formed became smaller and smaller; on the day of departure, the participants hurried up homewards, mostly without greetings, some at the beginning of a considerable but short career—what they could only suspected because of the military noise. Because the Solidarność was growing stronger and stronger, General Jaruzelski imposed a war law at the end of the following

G. Rüdiger, *The Astrophysical Observatory Potsdam - Triumph and Tragedies*, Astronomers' Universe, https://doi.org/10.1007/978-3-032-06294-9_8

Fig. 8.1 Illegal poster in Warsaw memorising the Gdansk Agreement of 31 August 1980. The political labour organisation Solidarność united 10 million members

year with the odd justification that otherwise the socialist brother nations would invade (Fig. 8.1). Because of the allegedly desolate Polish economic situation, the GDR school-pupils and their parents packed Christmas parcels for the "hungry children of Poland". The increased military activity on the German side of the river was only lifted again at the end of the war law in Poland, Solidarność had survived this period completely unimpressed. After official visits to Polish academy institutions had again become possible—with a special permission from the General Secretary of the AdW—I travelled with Tschäpe and Elstner to the astronomers Smak and Dziembowski from the Copernicus centre of the Polish Academy in Warsaw, which had been built with U.S. money including computer centre and guest house. At the suggestion of our director Schmidt we had formed an accretion disk working group in March 1986. Gas is collected in flat disks around young stars or black holes, whose potential energy is converted by friction into radiation which can prove to be extremely intense. The cause of the necessarily enormous viscosity—pointing to the existence of turbulent flows in the disk—was still an open question. Although the (magnetic) solution to the problem was actually obvious, we did not find it ourselves then or later. We checked the instability of Kepler disks as due to sound waves, nonaxial symmetric or nonlinear influences, all ideas without success.

At Warsaw Central Station, Tschäpe walked towards one of the numerous illegal taxis, which we had been warned not to use because of illegal activities. He abruptly asked the driver for the grave of Pastor Popiełuszko who had been killed by secret service officers a year earlier and whose funeral had been attended by more than 800,000 people. It was already too late in the evening to go to the cemetery but an appointment was made for the next day.

In his office, director Dziembowski told his amused visitors that he watches the revival of his science during the day and the decay of communism in the evening news. During a private dinner with Kazik Stępień from the Warsaw University Observatory, kilos of printed matter from Solidarność were passed around. This political movement would soon become so strong that Dziembowski's prediction would also be realised in Poland. We gave our host a new manuscript for the Polish journal Acta Astronomica and a long German salami. The publication is now forgotten but the sausage is not.

Immediately after return to Potsdam, a report on the political situation in Poland was due for the AdW administration in which I described in detail the story I had heard somewhere about Polish farmers who urgently demanded fittings for their horse-drawn buggies. The party secretary looked at me blankly for a long time after reading the text. The other piece of information, that in Poland members of the socialistic party hardly ever become professors in the physical sciences, was kept secret to him, as was the calm self-confidence with which the people of Warsaw looked at their royal palace in reconstruction and the old town houses on the market square. No Polish person had explicitly pointed out this national world-wide wonder to us.

An Observatory Gets Lost

Krause had managed to organise a meeting "Stellar and planetary magnetic fields" on the Telegraphenberg in 1983, at which the number of western lecturers far exceeded the number of talks by resident scientists (Fig. 8.2). Beforehand, he had to certify that each of the visitors was totally progressive,

Fig. 8.2 "Stellar and planetary magnetic fields", Potsdam-Telegraphenberg 1983; left to right: Gußmann, Krüger, Wiedemann, Bräuer, Rädler, Galloway, Greiner-Mai, Mann, Tuominen, Krause, Roberts, Fuchs, Oetken, Hubrig, Gerth, Ruzmaikin, Rüdiger, Belvedere, Paternò, Stępień, Stix, Cupal, Mestel, Schmidt, Staude, Franck, Schöneich, Webers, Jochmann (incompl.). (Photo H. Strohbusch. Courtesy of E. Kühmstedt nee Strohbusch)

peace-loving and open to international recognition of the GDR—which he did, bashfully, on his own. The lectures were published in the Astronomische Nachrichten as Schwabe's first communication on the sunspot cycle had also appeared there. Stix,[1] Mestel[2] and Stępień[3] set the focus of the conference, accompanied by presentations of the latest results of almost all Potsdam spectroscopists and theorists. In contrast, the head of photometry, Schöneich, only appeared in the conference photo; despite a prominent audience there was no contribution from him and his group, even at its best times the AOP had already disintegrated in terms of content and politics.

At the conference Ilkka Tuominen from Helsinki had explained his plan to interpret the recently discovered temporal fluctuations in the solar rotation law[4] as a consequence of the oscillating magnetic fields causing the spot cycle in order to obtain data for testing the dynamo models. He had come to link the equations of magnetic field generation and those of the mean flows in the turbulent convection zone of the sun. He wanted to incorporate the calculation scheme for the inner solar rotation into his code. For this project, Ilkka was ready to come to Potsdam very often because I could not travel to Helsinki

---

[1] Stix (1984).
[2] Mestel (1984).
[3] Stępień (1984).
[4] Howard & LaBonte (1980).

to do the job there. I was only allowed to attend science meetings in the Eastern Bloc with Western participation under control of the head of delegation. Ruben reported on his periodic visits in Helsinki to the Academy administration: *There is a travelling cadre available for the contacts (Prof. Ruben, Prof Krause), which are organised annually in both directions.* Relations with Finland were based on an agreement between the two academies. According to a surviving document, the traveller was instructed by the MfS to conduct military espionage in Finland.[5] The primacy of science only played a minor role in the Academy of Sciences of GDR.

In August 1980, the COSPAR workshop organised by Paul Roberts in Budapest enabled first face-to-face encounters with researchers from West Germany (Fig. 8.3). I heard the new ideas of M. Stix about the currently detected 5-min-oscillations of the sun revealing the inner rotation law. Together with F. Busse from Bayreuth, we went on a run to the Hungarian National Museum to see the exhibition of St Stephen's Crown guarded by heavily armed soldiers, which had just been sent by the United States to the Hungarian government on the condition that it must be displayed in public.

After my talk about the Lambda effect, Roberts suggested I write a book on the large-scale flows arising in rotating stellar convection zones for his

Fig. 8.3 Budapest 1980; left to right: Soward, Pudritz, Nightingale, Proctor, Muth, Weiss, Rädler, Childress, Jones, Roberts, Kerridge, Rüdiger, Loper, Stix, Ruzmaikin, Moffatt, Galloway, Léorat, Insertis, Krause, Acheson. (Photo F. Busse. Courtesy of F. Busse)

---

[5] Appendix 9.

monograph series The Fluid mechanics of Astrophysics and Geophysics with the working title "The Differential Solar Rotation". I didn't comment on it during the meeting. Back in Potsdam, I carefully began to study the old exciting discoveries and ineffective attempts to explain the observations and only then the offer was accepted. The observers, their results and some early theoretical approaches later filled the first two chapters of the monograph.[6] During the remaining time on the Telegraphenberg I was able to organize about half of the planned material. After several crises—the basic turbulence-induced Lambda effect had hardly been discussed anywhere so far—the complete manuscript was submitted to the Akademie-Verlag in June 1985 and a year later sent to Roberts in Newcastle for edition. After all, the book was published simultaneously by the Akademie-Verlag Berlin and Gordon and Breach Science Publishers in London in the early summer of 1989. I was told that, according to GDR law the title was exclusive property of the Akademie-Verlag and that the version based on the design of Roberts' series had to be sold under licence. The East-Berlin publishers had probably just wanted to improve their valuta balance. The publication of the book was therefore delayed by years, and one can imagine what difficulties would have had to be overcome without the authority of a P. H. Roberts in Newcastle/England. It might be that the typewritten manuscript, produced with two carbon copies, would have become lost in the chaos of the dissolution of the Akademie-Verlag— just as happened to all my documents relating to a dispute with Treder about an article for the public institute bord, sent to the Academy magazine "spectrum" without a copy at the beginning of 1990.

In Budapest, R. Pudritz presented a novel application of the dynamo theory. Accretion disks rotate like planetary systems according to Kepler's law. However, they are not spherical like stars or planets nor are they elliptical, but flat with an outwardly increasing thickness. It is no longer possible to find strict boundary conditions for the Maxwell equations for such geometry. Pudritz had entered the mathematically rich field of approximated boundary conditions for the first time and found later many imitators, including us.

The following year, a high-level IAU colloquium on Problems of Solar and Stellar Oscillations was held in Crimea, which a Potsdam delegation with Kurths, Rädler, Rüdiger and Staude was able to attend. The main topic was actually supposed to be the 160-minute oscillation of the sun which was observed there under his well-known and active director Severny with his assistents Tsap and Kotov. More lasting was to meet Ludwig Deubner as one of the leading helioseismologists at his time. Together with his collaborators,

---

[6] Rüdiger (1989).

he had spectacularly found from analyses of the solar 5-minute oscillations that the linear rotational velocity increases down to around 15,000 km below the solar surface.[7] As this quantity should decrease for uniform angular velocity, the conclusion is that the actual angular velocity of the sun increases inwards, a very basic and persistent result. It is known today that the angular velocity below the equator indeed first rises considerably, then reaches a broad plateau before falling sharply again deeper in the solar convection zone. The resulting difficulties for the construction of solar dynamos later became known as the "dynamo dilemma". Deubner—who had noticed my enthusiasm about his results—and I said goodbye as emotionally as if it were fixed that the Iron Curtain would exist for hundred years. After just 10 years, Ukraine had become an independent state. Then I met Deubner as a three months guest of Harold Yorke in Würzburg, where I also saw Michael Kaisig and Hans Zinnecker. The first numerical simulations of the electromotive force for an ensemble of supernova explosions in rotating magnetised galaxies were started here by Kaisig[8] continued by Oliver Gressel later in Potsdam.[9] In Crimea I also met M. Knölker while swimming in the Black Sea. His matrix method[10] for determining the natural frequencies of solar oscillations was so convincing that, under normal conditions, I would have immediately travelled to Göttingen or Freiburg to join in the calculations, or at least for discussions. From Potsdam via Göttingen to Freiburg, the route that the friends Mattig and Schröter had taken in due time and forever.

The return of the Astrophysical Observatory to the international conference calendar, however, was not allowed to last after 1983 (Fig. 8.4). By the end of this year, we were said to pack our bags to leave the Telegraphenberg and to move to the Sternwarte Babelsberg into a cheap functional building called White House, made of iron and white cardboard, originally built as a residence for guests of all Potsdam academy institutes. When we worked there, the house accommodated astronomers on the lower two floors and sleeping guests on the upper floor, all rooms equipped with washbasins and showers with the typical plastic valve handles of the GDR.

Most astronomers had protested without success against their transfer to Babelsberg; only the spectroscopy group was finally allowed to remain in the refractor building because of its immobile measuring machines, while the powerful technical laboratory of H.-J. Hubrig were completely thrown

---

[7] Deubner et al. (1979).
[8] Kaisig et al. (1993).
[9] Gressel et al. (2008).
[10] Knölker & Stix (1983).

Fig. 8.4 Astrophysical observatory, last group photo 1983; left to right: Jahn, Burghardt, Kempe, Schwarz, Orwert, Thomas, Röpke, Gerth, Haase, Friedrich, Želwanowa, Rüdiger, Stoof, Romstedt, Wiedemann, Klosa, Dyllong, Zander, Eschrich, Czeschka, Witte, Berkholz, Schilbach, Krause, Werder, Bischof, Berlin, Meyer, Domke, Hempelmann, Röpke, Haubold, Strohbusch, R.-D. Scholz, Straeck, Nix, Dick, Gußmann, Wittke, Neuendorf, Hildebrandt, Hirte, Töpfer, Scholz, Schöneich, Hubrig, Nader, Rädler, Falge, Tschäpe, Ruben (Missing: S. Hubrig, Lange, Oetken, Stahlberg). (Photo H. Strohbusch. Courtesy of E. Kühmstedt née Strohbusch)

out of the ZIAP altogether. This spin-off was part of the shift from basic research to technical equipment construction for INTERKOSMOS satellites.[11] The vacated workrooms on the ground floor of the original main building now belonged to the new, rapidly growing technology group which had to work from now on for the entire Department Geo- and Cosmos Sciences; the upper floor and the complete former library house went to geophysics groups of ZIPE. The time of the Astrophysical Observatory on Telegraphenberg had come to an end despite all later attempts to revitalise it. According to the official language regulation the name was to remain so that the observatory had formally not been closed though it had disappeared. Even the planned transformation of the refractor building after the successful reconstruction of the instrument and the dome hall in 2006, together with both the Einstein and Helmert Towers into a country-wide museum was lost in competence wrangling. The spirit of the Observatory, Marie-Luise Strohbusch, moved with her family from the top of the Telegraphenberg to a more comfortable apartment in Potsdam downtown only when the last revival fanfare had died down—after decades of carrying food for her family up the mountain in her rucksack.

---

[11] Buthmann (2020) p. 913.

Magnetic Galaxies

In October 1983, a magazine article[12] ended with the statement that the author's utensils and those of his colleagues from the Astrophysical Observatory are packed and awaiting transport from Telegraphenberg to Babelsberg. The history of the observatory had lasted almost 110 years, years that were also dedicated to the physics of the sun. In 1941, at our new domicile, the Sternwarte Babelsberg, Ludwig Biermann had explained far in advance that sunspots are dark because the enormous magnetic field prevents the convective transport of heat. The almost apolitical magnetic-field Section had not survived the move to Babelsberg as a whole. The spectroscopy group was left on the Telegraphenberg and the photometry group moved with flying colours to a new section headed by Ruben. The dynamo group was suddenly on its own and its will to look for astronomical observations was in danger of being forgotten. They'd rather study their equations than the sky, a long-established astronomer in Babelsberg had mockingly greeted. He was right in the sense that indeed one wants to measure the magnetic fields of cosmical objects and another one wants to know the equations that contain these fields as solutions—that's how physics works.

For a while it turned out better than anticipated. The well-respected Karl-Heinz Schmidt became new director at this time, the mathematician Detlef Elstner joined the dynamo group in 1984 and in 1986 the physicist Reinhard Meinel arrived, all from Jena. When in 1985 U. Dyllong who was formally assigned to the MHD group, left the ZIAP and the GDR, Meinel applied for the vacant position. Once Dyllong asked for the printed version of the Final Act of the 1975 Helsinki Conference on Security and Cooperation in Europe, which demanded that all participating states—including GDR—respect human rights. On this basis, he successfully justified his application to leave this country because of family reunification. Unlike other applicants, he was allowed to stay in the institute until the day of his expatriation. On the last night, he said goodbye to me, my wife and his long-term girlfriend in his former apartment.

Schmidt had encouraged Meinel to come to Babelsberg years earlier at the Jena observatory, in case that a position becomes free. The former student of a Mathematics Special School, winner of the Physics Olympiad and research student at the university was indeed given by Schmidt the position in the dynamo group. Despite his striking early successes, Meinel had first to buy his place in the physics class in university with a commitment to three years of

---

[12] Rüdiger (1984).

military service, which led to two lonely escape attempts, one before and one during the army time. He was then released from the army, obliged to keep this period secret for the rest of his life, and was finally able to begin studying physics and mathematics in the current first semester of the Friedrich Schiller University Jena.

Together with Dyllong I had to look through the written legacy of Wempe who died at the end of 1980. There were the very different denazification certificates for astrophysics and geophysics on the Telegraphenberg, a strange proposal by Wempe to the Soviet Army to explore rocket launch sites on the moon with the Great Refractor, but also documents about the assistant Helga Starke who had fallen into the hands of the Russian military courts in 1951/52 for a trivial accusation. What an experience that I was able to hand over Wempe's documents to her at a meeting in our private home shortly before the world-wide pandemic starting 2020.

Supported by the comfort of a hotel room on the upper floor of the White House, an own office on the ground floor and close private connections, Ilkka Tuominen became a permanent visitor in Potsdam (Fig. 8.5). Originally a meteorologist, he was introduced to astronomy by his father who was well-known in the solar research community. Ilkka began with sunspot statistics and spectroscopy of magnetic stars before calculating the internal structure of

Fig. 8.5   Left to right: Ilkka Tuominen (1939–2011); Axel Brandenburg

rotating stars at the Moscow Astro Soviet in the early 1970s. Although this adventure remained an episode, his lifelong association with many Ukrainian and Russian observational astronomers did not. Together with friends from the Crimean Observatory, he would later design and build the high-resolution spectrograph SOFIN, which has been in operation at the Nordic Optic Telescope on La Palma since the early 1990s and has led to highly acclaimed results.[13]

It is possible that the construct of Finnish-East German cooperation also originated from Tuominen's Moscow period. Perhaps he occurred on this ticket first in 1981 with a lecture on rotating stars in Potsdam, surprised by the subsequent detailed discussion on the Zeipel paradox and the equator darkening of rotating stars as well as our nightly conversation about the Tarkovsky-movies Rubljov and Solaris in the highest-located nightclub in Potsdam. Since this visit, his professional interests have focussed entirely on turbulence and magnetic fields. He will later form a worldwide network for cool star research at the Helsinki observatory. Our mutual dream, much discussed in Potsdam, of organising an IAU colloquium The Sun and Cool Stars: Activity, Magnetism, Dynamos in Helsinki in 1990 on the occasion of a solar eclipse perfectly visible in Finland—also to solve my travel problems—will come true, albeit in quite another way than anticipated.

During the late 1980s, the Soviet space programme became more interesting for astronomers. The two Phobos missions to Mars and its moons, launched in July 1988, revealed that planetologists and specialized researchers for structure and formation of planetary systems did not exist in GDR. Even more dramatic were the challenges posed by the plans for the Spectrum-X-Gamma mission with the Soviet-Danish X-ray telescope SODART, which was imagined to be launched monitoring extra-galactic targets for mid-1993. Also here was little or no expertise among the GDR astronomers. The Institute for Cosmos Research (IKF) in East-Berlin, which was newly founded in 1981, was therefore given a special section of Extraterrestrial Physics with 35 scientists in groups for planetology, cosmic plasma physics and extra-terrestrial astronomy, whose work was primarily intended to fit in with the Soviet INTERKOSMOS programme. Secrecy and recruitment requirements were directed by internal rules issued by the GDR government specifically for the IKF. Only the planetary research group has later been recommended by the German Science and Humanities Council (WR) for continuation at a DLR successor institute for Planetary Remote Sensing.

---

[13] Berdyugina & Tuominen (1998).

The activities of the new IKF research section had only minor impacts on the activity of the ZIAP. J. Greiner (IKF) and W. Wenzel (Sternwarte Sonneberg) searched extensively on the Sonneberg sky monitoring plates (since 1928) for optical counterparts of the gamma-ray bursts (GRB) occasionally registered by satellites which last a few minutes at most. After reviewing thousands of photographic plates, an optical burst was actually found for 14/15 August 1966—simultaneously on 3 plates—within the error range of the position of a GRB on 6 October 1978. In order to be able to exclude local light phenomena in the sky such as planes or meteors, corresponding images from other observatories would have been required but without success.[14]

Another interaction concerned the planetary research group of the IKF. From around 1984, a concept of self-gravity to explain the Titius-Bode series of the planet distances was developed. Gravitational instability in flat disks provided a system of rings that were believed to contract into isolated planetary bodies resulting from an unknown azimuthal instability.[15] This exciting suggestion triggered own calculations. Did the instability really exist as a nonaxisymmetric mode or were the assumptions used too idealistic? However, the solar system was not a system of narrow rings but the distances of the planets increase with the size of the distance from the sun. This proved to be mathematically possible only if the disk mass—more than the mass of all planets—were greater than a tenth of the solar mass, which probably it never was. Schmidt had the thankless task of reviewing a book by D. Möhlmann on this subject for the Akademie-Verlag, which he partly delegated to me. He was certainly not amused about my report when I told him about our own nonlocal formulation—it was a diplomatically difficult challenge of unclear consequences. I never saw Schmidt's certainly cautious report, but neither did Möhlmann's book, not until today.

Stars and planets have well-defined surfaces. With sufficiently fast rotation, magnetic fields can be generated in the turbulent interior of the spheres which can easily be matched to moderately far-reaching current-free fields in the outer space. Seen from a large distance, there is a turbulent clump that is strongly magnetic inside and slightly magnetic outside. It will be possible to imagine that the star is a bubble located in a distant envelope of a field-free material. For numerical solutions, the inner bubble does not have to be spherical and sharply defined, but can also look, e.g. like a flat galactic disk. The turbulence-defining variables and the rotation laws are only defined by their isolines of any form, and the cosmic objects may also gradually migrate into

---

[14] Greiner et al. (1991).
[15] Möhlmann (1984).

their surroundings. This was the philosophy behind our new dynamo code, which was developed for use on galaxies in Babelsberg and programmed by Elstner in just a few weeks.[16] At that time, a top-secret western PULSAR computer with 32-bit architecture had been installed for military purposes in the Sternwarte, which had to be operated and guarded day and night. Our computing programme also contained numerical innovations that made it possible to obtain steady-state solutions for a single galaxy after just a few days of computing. Models developed elsewhere, so-called flat-disk dynamos or those in ellipsoids, had comparatively significant disadvantages. These constructions often proved to be mathematically too demanding,[17] while it even remained unclear whether the frequently used disk dynamos—infinite in the horizontal directions—really always provided the only interesting stable solutions.[18]

Later, single stars or accretion disks were also treated according to our concept "without sharp boundaries", but the trigger for its rapid development was a major astronomical event. We had heard that the 100-m-radiotelescope of the Max-Planck-Institute for Radio Astronomy in Bonn had succeeded in visualising extensive and beautiful magnetic structures in galaxies (Fig. 8.6). Krause—via Rainer Beck already in contact with MPG-director Wielebinski— was excited: We can now see the toroidal magnetic field belts that we have been talking about for decades without ever having seen them. Galaxies are reasonably transparent; in addition to the inner magnetic fields their inner rotational behaviour is also accessible for direct measurements unlike stars. But there was a complication for the theory. While R. Beck had derived a very regular, axisymmetric magnetic structure for the Andromeda nebula M31, his colleague Marita Krause presented a two-armed nonaxisymmetric structure for the galaxy M81.[19] Are there really axisymmetric celestial bodies that preferentially induce nonaxisymmetric fields despite the action of differential rotation? According to the Cowling theorem of 1932, no axisymmetric magnetic fields can be maintained by any laminar flow system. The regular magnetic fields of the Andromeda nebula therefore require the existence of a turbulence-driven dynamo. But how can M. Krause's differing observations be understood in this context?

---

[16] Elstner et al. (1990).

[17] Stix (1975).

[18] Rädler & Bräuer (1987).

[19] Beck (1982), Krause et al. (1989).

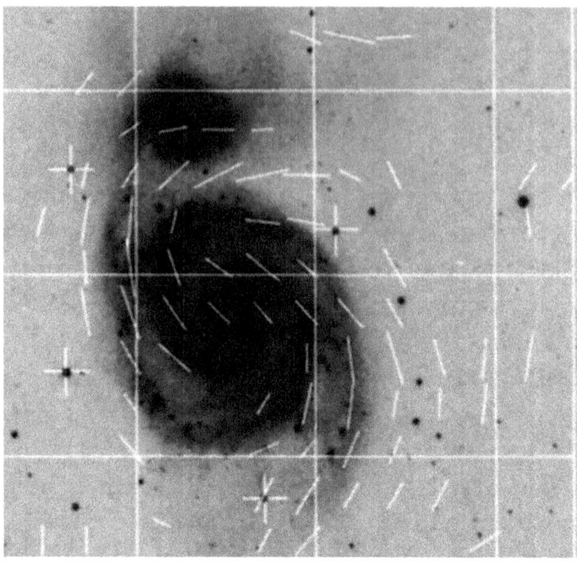

Fig. 8.6 Magnetic field lines in M51. One of the first registration of large-scale toroidal field components in galaxies. Beck (1987)

In 1988 a Moscow group stated in their book[20] that any nonaxisymmetric field in the Andromeda nebula should decay according to their calculations, the field should grow, however, in M51, albeit much more slowly than the axisymmetric field. The existence of nonaxisymmetric magnetic fields in galaxies was thus claimed as clarified. Meinel, however, argued, that only the solution with the minimal eigenvalue is relevant and this is always the one for the axisymmetric fields. Only this solution remains stable in the transition to realistic nonlinear models while all other modes do not. Months of computing time to determine the higher eigenvalues turned out to be needless. In fact, later nonlinear calculations have never produced a bisymmetric field structure,[21] whose existence is still not understood for galaxies at least if one ignores the spiral arms.

Enough material for disputes; Krause and Wielebinski wanted a discussion meeting. As the young active members of the Potsdam dynamo group could not travel to Effelsberg to meet the radio astronomers, Krause decided to organise the meeting at the Sternwarte Babelsberg without consulting the academy administration in East-Berlin. Richard Wielebinski with his Australian passport was sceptical; he only knew the GDR as an annoying

---

[20] Ruzmaikin et al. (1988).
[21] Elstner et al. (1992).

transit country on his journeys between Bonn and Poland. In the Introduction written in October 1989, Krause gingerly spoke of a workshop in preparation for the following IAU symposium in Heidelberg. But the sensation happened: one day in autumn 1988, an eastern minibus drove to the Dreilinden border crossing to pick up the director of one of the world's largest radio telescopes and his closest colleagues among them Beck, Kronberg and Marita Krause to take them to the guest house in Babelsberg.

The giant instrument in Effelsberg had been planned by O. Hachenberg, former director of the Heinrich Hertz Institute (HHI) in Berlin-Adlershof, with the support of Krupp and MAN, after he had moved to Bonn during the installation of the Berlin Wall. In 1970, in the middle of the construction phase in Effelsberg, the Polish-born radio specialist R. Wielebinski from the University of Sidney, became Director of the MPG institute for Radio Astronomy in Bonn. The MPG institute absorbed the activities of an older university instititute of radio astronomy to build a 90-m-mirror under the leadership of Friedrich Becker who in the 1920s established a satellite station of the AOP in La Paz (Bolivia).

I was allowed to pick up the Helsinki group with Tuominen, Donner and the doctoral student Axel Brandenburg from Schönefeld airport for the conference with my minicar Trabant, driving far south around Berlin, with a beautiful view of the white West-Berlin Gropiusstadt and with less beautiful impressions of the grey, decaying eastern town Teltow. That western visitors may only be hosted by authorised persons did indeed hold in the AdW but was largely ignored.[22] The two Finns kept quiet about my uninterrupted German talking with the young Axel who just nine months later will show me the most beautiful places around his hometown Heide in the North Sea wind. The Moscow group with Shukurov and Sokoloff had been transported by Rädler with his even smaller car probably on the same route.

Beck was the first conference speaker presenting his statements: The Effelsberg radio telescope has provided polarisation data for twelve galaxies. M31 and IC342 are dominated by axisymmetric structures, while others probably have a nonaxisymmetric magnetic field structure. Dynamo calculations for axisymmetric gas discs always provide the fastest growth for the lowest axisymmetric magnetic field mode. That's all we know in Bonn: theorists, now fight for the correct explanation (Fig. 8.7)!

---

[22] K. H. Moffatt in his condolence to Irmgard Krause dated 25 March 2024: "Fritz took me on a mushroom-hunting expedition in the fields near your home... I stayed in Potsdam for several days, and was entertained also by Karl-Heinz and by Günther at their homes."

Fig. 8.7 "Magnetic fields in Galaxies", Potsdam-Babelsberg 1988; left to right: Shukurov, Gvaramadze, Rädler, Fröhlich, Elstner, Eschrich, Jansen, Gerth, Meinel, Sokoloff, Minor, Tuominen, Beck, Harnett, Brandenburg, Sokhadze, Seehafer, M. Krause, Tosa, Chagelishvili, Moss, Wiebicke, Kronberg, Donner, Rüdiger, F. Krause, Fujimoto, Geppert, Wiedemann, Buczilowski, Wielebinski, Kliem. (Photo R. Beck. Courtesy of R. Beck)

Behind the Watchtower

All talks of the galaxy meeting took place in the small lecture theatre of the Sternwarte to which the external participants from the guest house had a much shorter way than the locals from the Sternwarte. In May 1987, the majority of the institute's theorists had been moved overnight from Babelsberg to a magnificent villa[23] in the Stubenrauchstraße on the shoreline of the Griebnitzsee. The brick building with the huge sundial on the street side, wall fountains and a large garden, full of natural stone-walled rose walkways and a huge copper beech high above the Griebnitzsee, had the particularity of being located directly on the GDR border fortifications (Fig. 8.8). Moreover, there was the same concrete structure with 4 m height with a slightly narrower death strip on the street side and at the end of this street 50 m ahead. The houses behind it already belonged to West-Berlin. The course of the Wall

---

[23] Also Siemens villa or country house Koettgen, built in the mid-1930s. The garden is considered exemplary national-socialist with a fountain, vine pergola and copper beech.

Fig. 8.8 Stubenrauchstraße 26, from April 1987 an exotic laboratorium for theoretical research groups (dynamo/accretion theory, cosmology, magnetosphere/plasma physics) on a waterfront property fenced in on three sides by the GDR border fortifications

carefully followed the old Berlin city maps from the pre-war period. The head of the plasma group[24] had claimed a room with a view to the lake to make it more difficult for Western secret services to spy on his highly important results from the West-Berlin side of the street (Fig. 8.9). Border guards on a watchtower controlled the access to this artificial fortress between Potsdam and West-Berlin, which one was only allowed to enter with a special ID card (Fig. 8.10). Mostly the border soldiers remained inactive, trusting that the official authorities would carry out careful security checks on the members of the astronomer group. In GDR times the villa was a former guest house of the film company DEFA, whose administration had probably become too nervous by the permanent complications in the closed border area. The vice director of the ZIAP acted as the house manager, usually he only attended the weekly seminars. The rules were extensive, with a strict photo ban at the top of the list, which was only ignored a few days before the fall of the Berlin Wall. There was only a single telephone apparatus with one or two extensions and a

---

[24] In 1984/1985, a large group for magnetospheric/ionospheric geophysics joined the ZIAP headed by C.-U. Wagner (1935 - 1989).

Fig. 8.9 The Berlin Wall surrounded the villa Stubenrauchstraße 26. Note the watchtower in the background as the entrance to this astronomy fortress. (Photo P. Rohn. Courtesy of P. Rohn)

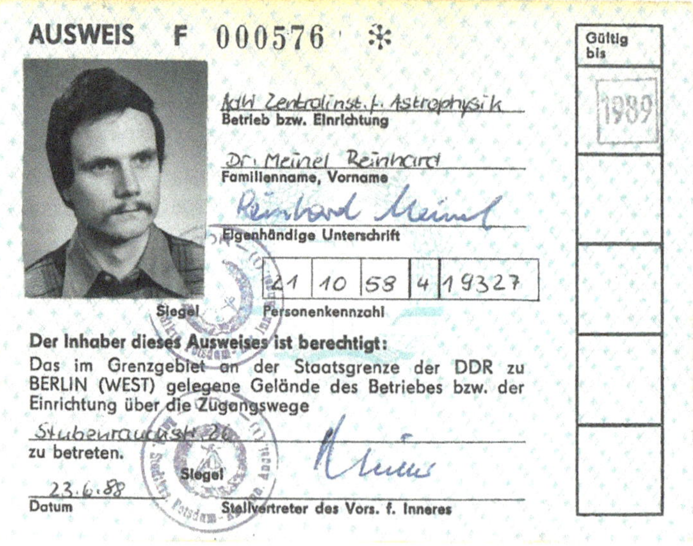

AUSWEIS   F   000576   ✳

Gültig bis

1989

Adw Zentralinst. f. Astrophysik
Betrieb bzw. Einrichtung

Dr. Meinel Reinhard
Familienname, Vorname

Reinhard Meinel
Eigenhändige Unterschrift

21 | 10 | 58 | 4 | 19327
Siegel   Personenkennzahl

Der Inhaber dieses Ausweises ist berechtigt:

Das im Grenzgebiet an der Staatsgrenze der DDR zu BERLIN (WEST) gelegene Gelände des Betriebes bzw. der Einrichtung über die Zugangswege

Stubenrauchstr. 26
zu betreten.

23.6.88
Datum

Siegel

Stellvertreter des Vors. f. Inneres

Fig. 8.10 Border pass for the frontier area Stubenrauchstraße from June 1988. Only a final extension of the document for 1989 occurred. Courtesy of R. Meinel

big bell that frequently rang through the house for everyone to hear. At some point, two simple 16-bit ROBOTRON 7100 computers were set up, each with two floppy disk drives, and the people were soon queuing up in front of them to do their paperwork; it was only at this time that weekend work was tolerated in the border fortress. Some users had put together their own sub-programmes to display Greek letters by the dot matrix printer. The number of disks was strictly limited and the disk distribution were managed from the onset by Detlef Elstner who later headed the IT commission of the future Astrophysical Institute Potsdam for many years. An electronic connection line between villa and Sternwarte, which was initially quite slow, was soon set up and gradually improved under his aegis. The move to the borderland fortress had interrupted old working structures; the few survivors remaining from the old large MHD section now competed in two separate groups to calculate galactic magnetic fields. There were no new connections with the observers remaining on Telegraphenberg, and only one common project on magnetic stars survived. The newly formed accretion disk group also found itself separated, Tschäpe and Fröhlich were not allowed to join the rest of the group in the villa. Our calculations of turbulent heat transport and its effect on the stability of accretion disks were mostly coupled together by the green telephone monster.

At the beginning of 1988, one or two of my inner batteries had run out. In retrospect, symptons were already recognisable much earlier during the final production of the monograph about solar rotation, when the collection of the extended bibliography was postponed from week to week until Dyllong who was waiting for his final departure, took pity on me and compiled the numerous index cards ready for printing within a few weeks. At the clinic for the treatment of such diseases in East-Berlin, I underwent six weeks of group therapy with relaxation exercises, sport and lots of simple housework. Political conflicts were generally not taken seriously by the numerous therapists because of "you can't change that and we can't change that; get used to it". Being part of an international scientific community as an isolated GDR citizen was not on their list of treatable constellations. The question of whether this my business construction was stable and how long it could last maximally could not be asked, not even in individual interviews. Only later did it become public that three of the about ten therapists had an IM status of the MfS. After all, the six-week break was enough to restore the internal forces. When I later took out a life insurance policy, there were problems because of this medical episode as the Western insurance agents were suspicious of my time in the apparently notorious institution.

The following summer, the staff at Stubenrauchstraße pushed ahead with the translation from Russian into German language of the comprehensive Soviet encyclopaedia Physics of the Cosmos, which Liebscher had negotiated with the Akademie-Verlag. I had to edit the interesting chapter on X-ray astrophysics together with Svetlana Hubrig as a native speaker and delivered it on time but we never heard from the project again. The finished translation was handed over to the publisher, it was said, and then probably disappeared, lost in translation.

Late in the evening on 9 November 1988, candles flickered in empty canning jars on the ground in front of the building next to the main post office of Potsdam. This building stands on the footprint of the old synagogue, and had a young medallion attached to the wall in preparation for a hoped-for visit to the USA by Honecker.[25] Deeply moved, I got a candle from my home, drove back to the post office building, left the engine running and, after getting out, opened the car door, lit the light and quickly disappeared by careful observing the surroundings. A quiet excursion into the private political feeling, just one year before the next historic November 9. Until then, I would meet the head of the tiny protestant Training Centre for Community Education as one of the candle activists, also as he had the only freely usable modern photocopier in Potsdam. Even when I spontaneously and carelessly wrote a protest letter to a journalist of a daily newspaper in favour of Israel, which was regularly accused by the official media, there was no any access to a photocopier hence the letter about the early Palestinian conflict has been lost. Antifascism as its state doctrine had not automatically brought the GDR on the side of Israel, but on the contrary to one-sided support for the PLO. I would have named Göttingen, Boulder/Colorado and Tel Aviv in this ranking as my most urgent travel wishes—if someone had asked me but that never happened.

After the galaxy conference, L. L. Kitchatinov from the Siberian Institute of Terrestrial Magnetism, Ionosphere and Radio Waves in Irkutsk visited the dynamo group in Potsdam (Fig. 8.11). He was interested in turbulence issues of all kinds and had already published remarkable papers on the subject. He stayed with us for many weeks to prepare for his keynote lecture on solar rotation the following year at the IAU symposium in Kiev. He had brought with him first calculations on the magnetic influence on the turbulent transport of angular momentum in rotating flows. With such formulae, the temporal variation of the rotation law can easily be determined and compared with

---

[25] The surprising success to build a Mormon temple in Freiberg/Saxony in 1985 had a similar political background. Honecker was never invited to visit USA.

Fig. 8.11   Leonid Kitchatinov at Lake Baikal. (Courtesy of L.L. Kitchatinov)

observations. The result was dramatic: the rotation of the sun was more than ten times more strongly influenced by the magnetic suppression of the Lambda effect than by the global Lorentz force, which we had previously favoured with Tuominen. The old approach was thus already obsolete after a short time and did not find a continuation.[26] Three decades later, massive numerical simulations of rotating magnetoconvection will confirm the analytical results of our first German-Russian publication. When, in the course of Gorbachev's glasnost and perestroika, all scientists in Irkutsk were asked where they would first travel if they could, they wished for Cambridge or Stanford, Leonid—despite the amusement of his colleagues—nominated Potsdam at his first place. "Leonid, Potsdam belongs to East Germany" they all explained him ironically. A 20-year collaboration followed, which only ended when, due to age, I myself was only a guest at the institute and, unfortunatey, his application for my position was unsuccessful. He hasn't set foot in Potsdam since.

Only two of the later joint publications may be mentioned. In the 1990s, it was calculated that the very thin transition layer ("tachocline") in the rotation between the outer convection zone and the interior of the sun can be well understood if the latter is permeated from the beginning by a very weak large-scale magnetic field that has always been there and has nothing to do with the

---

[26] Rüdiger & Kitchatinov (1990).

11-year cycle of solar activity. For unknown reasons, the publication was not enthusiastically received by its reviewer and only appeared two years later, still early enough to initiate a long series of similar magnetic explanations world-wide.[27] Later, as a result of lengthy calculations on basis of the Lambda effect, we predicted that almost all main sequence stars should have approximately the same absolute difference in angular velocity between pole and equator which agreed well with the future observations.[28]

Why sunspots are dark and according to which rule they decay over days and weeks have always been pressing questions, the answers to which—as the first Potsdam publications basing on the concept of magnetically suppressed convection showed—could be used to draw conclusions about the influence of strong magnetic fields on turbulent plasmas. One key question, whether the dark area of the spot decays linearly or quadratically with time, could have been answered by analysing regular solar observations and measuring the temporal behaviour of the spot size. For decades, thousands of spot drawings had been made at the Einstein Tower, without being analysed for theoretical or other purposes, but less than hundred meters away from the Tower building we didn't know. The new approaches to radiation transport in turbulent stellar atmospheres fared similarly. Here J. Stahlberg succeeded in finding a formulation based on the method of mean-field quantities in which the two known approximations "microturbulence" and "macroturbulence" are understood as natural limits for very small and very large turbulence elements. Freundlich's measurements in the Einstein Tower had provided the gravitational redshift only at the edge of the sun rather than in the centre before much later Schröter provided a first explanation including estimated turbulence conditions on the solar surface.

N. Seehafer moved to the Stubenrauchstraße from Telegraphenberg in early 1989. He had extrapolated photospheric magnetic measurements from active areas on the solar surface in the direction of the corona under the assumption of a vanishing Lorentz force ("force-free") and found that the resulting current helicity satisfies a surprisingly simple sign rule. In the northern hemisphere there were only negative values and in the southern hemisphere only positive values. Obviously, the difference was due to the Coriolis force of the global rotation; both the current helicity and the α effect are pseudoscalars with different signs in the two hemispheres. Based on a theoretical result from the literature, Seehafer had thus demonstrated the existence of a positive (negative) α effect on the northern (southern) hemisphere of the

---

[27] Rüdiger & Kitchatinov (1997).
[28] Kitchatinov & Rüdiger (1999), Reinhold et al. (2013).

sun, even with measurements from the Einstein Tower spectrograph.[29] Meinel who liked to emphasize that the existence of the α effect in rotating stars was written into his working contract, could feel relieved.

In June 1989, the IAU symposium on galactic magnetic fields took place in Heidelberg. Despite financial commitments from the MPG the authors of the new Potsdam dynamo concept remained locked out again. Krause had to present the most important results on their behalf.[30] Of course, the Academy presidents could have convinced the SED-leaders that an invited, externally funded lecture at a first-class congress was as good a reason for travelling as an athlete's participation in the finals of an international championship, an engagement on a foreign theatre or a visit to the birthday of an elderly family member living in West Germany. Even the cultural functionaries had successfully campaigned for the right to travel of countless writers, painters and filmmakers—the authorities of the Academy of Sciences formed the sad exception. It might be easy to find the reason. It was too risky for them as then professional excellence would soon have become a condition for almost all contacts abroad; a risk they preferred not to take as science was not so important to them. They all could not have the slightest idea that the system they had constructed over decades will not survive the year 1989.

---

[29] Seehafer (1990), Keinigs (1983).

[30] Krause et al. (1990).

9

The East German Revolution

Time for a New Forum

In the summer of 1989, the GDR security forces were probably already experiencing concentration problems due to overwork, when we were travelling back from a private visit to Flensburg via West-Berlin to the Dreilinden border crossing. To everyone's surprise, I had been allowed to accompany my wife to Flensburg for her mother's 70th birthday in August. Such DFA[1] trips, in addition to the escape stories of countless artists and writers—starting in 1977 with the spectacular emigration of the popular actor and singer Manfred Krug—were the dominant themes in the decaying GDR. We watched through the railway window the rapidly disappearing border fortifications convincing ourselves that we were really travelling into West. When we changed trains behind the border, the station mission offered coffee and bananas—never before had anyone gifted me anything at a railway station. We travelled on to Kiel in a railcar. "Did they make a monkey out of you with bananas?" screeched our cousin from the western peace movement in the evening when we told our relatives about the tropical fruits at the railway transfer station.

The next morning, I saw Matthias Steffen at the astronomical institute of Kiel University, whom I had already met in May during a tour of the Dnepr on the occasion of the highly impressive IAU conference "Solar Photosphere—Structure, Convection, Magnetic fields" in Kiev, organized by the tireless

---

[1] DFA = urgent family matters. Children had to stay at home as hostages. The AdW took only little notice of police authorisations for private west trips of its employees.

G. Rüdiger, *The Astrophysical Observatory Potsdam - Triumph and Tragedies*, Astronomers' Universe, https://doi.org/10.1007/978-3-032-06294-9_9

East-West mediator in solar physics J. O. Stenflo (Fig. 9.1). Here Ilkka Tuominen celebrated his 50th birthday with many friends of his global network from Cambridge MA, Crimea, Helsinki, Manchester, Moscow, Potsdam and Pulkovo. This conference was also unforgettable because of the frequent jokes about Gorbachev plus his wife and the opulent lunches which must have been the result of months of complex preparations. To top up my travelling funds, the institute library in Kiel bought the copy Nr. 1 of my brand-new monograph; with Steffen and his colleague Schönberner we later edited the Astronomische Nachrichten for many years in Potsdam together with K. Strassmeier as the editor in chief.

There was a paper kiosk in Flensburg where the SPIEGEL prophetically titled "Will the GDR explode? Mass exodus from Honecker's socialism." In fact, during our trip, the West German embassy in Prague began to fill up with fearless GDR refugees. Soon there were thousands, and on 23 August the diplomatic institution was closed to the public due to overcrowding, as was the Czechoslovakian border with Hungary at the same time. For several weeks, this embassy would be in the focus of international attention until 30 September, when Federal Foreign Minister Genscher in Prague announced from a balcony of the baroque palace, after first welcoming the thousands of refugees, that he had come "to inform you that today your departure …". Nobody knows the rest of this statement, it got lost in an exploding outcry. On 1 October, the first trains travelled undisturbed by GDR forces via Dresden to Hof but on 4 October there were already 6000 new refugees on

Fig. 9.1 Kiev 1989; first row, middle seat: Jan Olof Stenflo, President of the Scientific Organizing Committee

and in front of the embassy. In consequence, shortly before its 40th birthday on 7 October the GDR closed its borders in all directions.

For two days, Axel Brandenburg had picked us up in Flensburg with a rental car to tour North Friesland with its many spoken languages. We drove to his mother and his birthplace on the huge market square of city Heide for a crab meal and later travelled north along the main road to Seebüll with the Emil-Nolde-house. We saw the North Sea, marvelled at the new wind turbines and the countless European flags strongly flapping in wind. In front of most of the brick-built houses often situated on little fenceless hills, fruits were on sale. In the museum, Axel enquired about the tidal range in order to find persons from the region for smalltalk. In the evening in a Flensburg pub, over beer and fried eel from the house, he had buried the table top under notes and calculations, and the host waited patiently until midnight for us to finish our drinks and calculations.

"The sunspots will appear on 16 August," was the coded slogan for Willi Deinzer by a transborder telephone call to signal our planned arrival day in Göttingen. Krause had instructed him in this way at the Heidelberg conference in case permission to travel to Flensburg was granted. After a long train journey southwards through northern Germany in a comfortable saloon carriage, we arrived at Göttingen station and were recognised by Deinzer immediately, even though we had never met before. My seminar lecture on the solar dynamo and the internal solar rotation law took place in the afternoon, to my delight surrounded by old instruments in the Gauss room of the historic observatory including the famous 10-foot Herschel telescope made for Göttingen (Fig. 9.2). Schwarzschild had lived here before he came to Potsdam. I was so euphoric on this day that I only realised the hidden meaning of Dieter Schmitt's question about my opinion on the so-called dynamo dilemma days later in the train from Göttingen to West-Berlin. After the seminar, Prof Voigt wanted to make a note of the title of my lecture for the institute's annual AG report which would be published the following year. "No, sorry, I'm here illegally", I replied remembering the fate of Isolde Meinunger. "How can one illegally be in an astronomical observatory?" the professor shook his head. He personally took us to the Göttingen city cemetery with the graves of Max Planck, Max von Laue, Karl Schwarzschild and Ludwig Prandtl. I was also interested in Tobias Mayer, whose short tragic life I had recently described in a literary story. His observatory on a tower of Göttingen's city wall had been destroyed by a powder explosion during an observation night in the 7 Years' War. One such tower still stands or has been rebuilt near the observatory, and we visited it the next day. The original tower,

Fig. 9.2 Göttingen University Observatory, built in 1816, where Carl Friedrich Gauss (1777–1855) worked and lived. (Photo A. Wittmann, Courtesy of A. Wittmann)

however, was demolished in 1922.[2] In the evening at Deinzer's house we had a Silvaner wine in litre bottles from his favourite winery in Franconia. We had seen Göttingen and its observatory on family business and not as a triumph about the academic bureaucracy in GDR but nobody can take away the fact that we were in Göttingen, I thought on the way back, not yet realising that this was just the beginning.

I was able to read, in the files of the state security, my short contribution during the visit of Potsdam's Mayor at the Sternwarte Babelsberg on 17 April 1989 in preparation of the state-wide elections of communal councils at 7 May. It was about the absence of a green party in the GDR, not about doubts about the election system. From Monday 8 May, however, the whole republic was shaken by the discovery of counting fakes because the SED leadership had forgotten to abolish the public counting of the votes at the individual polling stations. For the first time, human rights activists, mostly from the evangelic church, had recorded numerous vote counts on the evening of 7 May and found drastic discrepancies with the published numbers next morning. They had registered 2192 dissenting votes in just 28 Potsdam polling stations, which had been reduced overnight to 1559 in the official election results for the entire city of Potsdam with about 100 polling stations! Complaints were dismissed, and recounts allegedly confirmed the accuracy of the results. On 1 November, just a few days before the border was opened, the

---

[2] Information A. Wittmann.

SED leader in the Potsdam city parliament required not to express distrust in the election commissions because the GDR always was a constitutional state. Kaminski—one of the undercover counters—shouted into the meeting that it's all lies after grabbing the microphone. The functionaries' class had obviously crossed a red line that it didn't even know existed. The common DNA of all the growing opposition groups in these days was therefore always the categorical demand for free elections.

At noon on Monday the corrected numbers of the communal elections in Potsdam had already spread around the Sternwarte. My conversation with Domke in the morning in his office had been recorded. It is expected, the institute director had stated at the staff meeting a week ago that ZIAP employees will fulfil their duty to vote already in the morning hours. The careless distributers of the real numbers—about 10% negative votes—were named at the regular SED group meeting on Monday evening. If they are not right, they will be fired, but if they are, we shall be fired, one participant was said to have formulated with intuition. Another person soon has been fired. Mayor Seidel was so careless to check the counting results in one public case and had to confirm the 78 NO votes (and 714 YES votes) claimed by Kaminski. Seidel's political career ended abruptly on 22 May because he didn't make the opposing votes disappear in time.

The summer of 1989 was a time of visible mass exodus. Already in May, Hungary switched off the electricity at its border facilities with Austria, after which up to 500 people crossed the porous border every day under the eyes of the soldiers and the international press, while the Hungarian authorities refused to intervene. 11 September was the day on which Hungary finally opened the border "due to technical and moral obsolescence", unimpressed by the desperate protests from East-Berlin. At the same time, in Stubenrauchstraße, after lunch and table tennis in a minor circle, relaxed and sweaty under the copper beech on Griebnitzsee, just a few metres above the border strip to West-Berlin, Meinel suddenly said concentrated with every word that yesterday he founded a new political party with Rudolf Tschäpe and a few others. I was sceptic but curious: what, how, where? "In Grünheide, in Robert Havemann's house." Krause, our third table tennis player, was suddenly in a hurry to get away to do some maths, he muttered. What's the name of the party? "Neues Forum, I don't want you to wait until the evening news. Not a real party, but political, will be officially registered next week."

In just a few days, the appeal formulated in Grünheide with the conclusion that the time has come—what indeed was prophetic—was circulated throughout the GDR, copied by hand, using typewriters, Ormig machines and the rare photocopiers in companies, schools and universities: "In our country, communication between the state and society has broken down. Evidence of this can be seen in the widespread disengagement and even withdrawal into private niches or mass emigration … We are therefore jointly forming a political platform for the entire GDR that will enable people from all professions, spheres of life, parties and groups to participate in the discussion and processing of social problems in this country." The 30 first signees from all sections of the civil society, from nurses to engineers, chose the maximally unpolitical name NEUES FORUM for their initiative. Shortly before end of August, West-Berlin's Mayor Momper, however, had announced that nothing could be achieved now by minor groups founding parties in the GDR. The SED "actually has the power and will retain it for the foreseeable future".[3]

Suddenly everything was different in the country, a secular voice in favour of the necessary reforms had risen. At first only a few people noticed, but in the end everyone did: the leading role of the SED was missing in the appeal, there was no longer any word of socialism, only the lack of democracy had been underlined. The authors had struggled for two days to fill in these blanks, and in the end, it was precisely this decision that led to the lifting of the mental bans that systematically had been erected for 40 years. The following year, millions of people took part in free elections for the first time in their lives, some of them with wet eyes.

Neues Forum's request for a legal status was vital for its supporters, but it came at a price. If I assume that your group stands on the ground of the constitution then you also accept article 1, which characterises the GDR as a socialist state and establishes the leading role of the SED, Meinel has been asked in an interview with a Potsdam journalist. His diurnal newspaper was the first one in GDR to open its columns to the new organisation. The constitution can be changed, Meinel replied, a leading party that is not legitimised by elections, this cannot work.[4] A party's claim to leadership clearly excludes democracy; it is this irresolvable contradiction that will finally lead to the end of the GDR. On 21 September, when the Minister declared Neues Forum as subversive, there were already 3000 supporting signatures.[5] An attempt to get our long-time ally Christa Wolf during a conspiratorial meeting in Babelsberg in favour of the Neues Forum failed because of the lack of a socialism clause in the texts. All other attempts to hire representatives of the established elite to join the new movements failed. Heiner Müller anticipated that "the resistance of intellectuals and artists who have been privileged for

---

[3]Tageszeitung, 22 October 1991.
[4]Brandenburger Neueste Nachrichten, 2 November 1989.
[5]Kukutz (2009).

decades … will achieve little if a dialogue with the long-silent majority does not happen".[6] From the end of November 1989, Christa Wolf tried to unite the interests of the SED and the opposition with her appeal "For our country" and collected one million signatures in favour of the "GDR as a socialist alternative to the FRG". Her campaign was heavily criticised, particularly in Saxony and its capital Dresden; Egon Krenz and a top MfS leader killed the whole project with their demonstrative signatures.

On 19 September 1989, Meinel and Tschäpe were presented individually at the Sternwarte to a triumvirate consisting of the new vice director of the Department Geo- and Cosmos Sciences, Hurtig, Ruben and a recorder. Ruben to Meinel: We have held you in high esteem up to now, but we are deeply disappointed at the way you are now doing the matter of counter-revolution. Four of the so-called signatories are employed at our Academy,[7] two of them in astrophysics alone, what a shame for our institute! The slightest misdeed or a questionable public appearance will have immense consequences for you. Meinel replied that he would continue to make sure that all political activities exclusively belong to the private time. In the plenum of the Stubenrauchstraße, Liebscher demanded Meinel to reveal his idea of socialism and got the answer that socialism doesn't work without democracy, that's the main idea. Liebscher advised him to withdraw his signature, then everything will soon be forgotten.[8] Just a few weeks later, on my request Liebscher had to write a letter to the Potsdam postal office to set up a telephone connection for Dr. Meinel in his private flat as soon as possible, because a great number of citizens (with an upward trend) are now seeking contact with Neues Forum. When I handed over this letter, the postal authority urgently wanted to know from me what will happen to his organisation in the future. Unfortunately, I couldn't anticipate that E. Schnell from the high-pressure laboratory on the Telegraphenberg who was organising his social-democratic demonstrations on Mondays, would be the next post minister of GDR. After only a few days Meinel indeed received one of the rare private telephones. News of this success soon got around, and a short time later I had to repeat the same procedure with parts of the Liebscher text for many more activists of the new political movement "Neues Forum" in Potsdam.

Academician Kautzleben as head of the Geo- and Cosmos Sciences Department of the AdW must have been very angry about the political machinations of the astronomers in Potsdam. He had convened the group leaders

---

[6] Neues Deutschland, 14.12.1989.
[7] R. Meinel, S. Pflugbeil, J. Reich, R. Tschäpe.
[8] Personal communication R. Meinel.

of ZIAP at the end of September, at which he appeared together with a supporting high-ranked functionary from the SED district administration to require to vote in favour of the final dismissal of Meinel and Tschäpe. There have been long speeches in which not much was said, but in the end nobody, not even the party members, signed the well-prepared sentence. The institute's party secretary at that time today remembers the general mood that was directed against Kautzleben. In a later vote of confidence of his own institute, Kautzleben only got 15%.[9] Nevertheless, the late Academy leaders succeeded in placing him as a member of a German-German evaluation commission for institutes in the field of geo- and cosmos sciences; only a letter of protest from the ZIAP dated 17 July 1990 to the GDR science minister and the chairman of the German Science and Humanities Council (Wissenschaftsrat, WR), Simon, stopped the sneaky strategy. Minister Terpe replied immediately that Kautzleben would not be included in the evaluation of the institutes in the Department of Geo- and Cosmos Sciences.

Four Interventions and a Funeral

After the violence against demonstrators in Berlin, Potsdam and other cities on 7 October 1989, I formulated together with the film director Rainer Simon a short statement beginning with: Who excludes or criminalises those seeking dialogue disturbs the development of our country, we stand up, therefore, for the official authorisation of the new political initiatives such as Neues Forum by the government. This rather mild position was signed dozens of times at the ZIAP, but not at all at the meeting of film directors and dramaturges that took place the next day.

On the morning of 18 October, a message appeared in the Sternwarte next to the entrance to the seminar room. The text consisted of a list of issues that worried me: the silence of the scientific elite, the hidden offer of a ROSAT project position in Munich and the inaccessibility of computers in West-Berlin or Göttingen. "Can anyone really imagine that people whose mail is opened, who have to put through cross-border telephone calls immediately to the director, who are forbidden to accept invitations … that such people can develop into Nobel Prize winners? Can we assume that those who create such an atmosphere want successful science at all?" Indeed, the number of all the talented scientists at the ZIAP—one could list them—showed no difference to the number in other European countries, one would only have had not to prevent

---

[9] Stark (1997).

their development. Neither the GDR government nor the upper academy/institute administration ever understood the crucial role of international collaboration for excellent science. The results of even voluntary and self-imposed isolation in connection with high motivation for science were clearly visible in the prominent case of the first director H.-J. Treder of the ZIAP.

The text ended with a list of seemingly utopian demands: Work visa for everybody, integration into the global computer network and free access to the Stubenrauchstraße location. Just 3 weeks later, after the fall of the Berlin Wall, almost all of these basics for our future work were fulfilled and foreign visitors soon started to proceed to the now free villa at the now free Griebnitzsee. Accidental encounters between visitors from different institutions sometimes did not go without comedy. Knölker and Schüssler from Freiburg unexpectedly met the Munich solar physicist H.U. Schmidt on the narrow staircase in the villa, minimal greeting ("Oh"), rigid faces. Knölker told later that it was as "before a table tennis match between North and South Korea" (Fig. 9.3).

To my list I had added an ironic comment about the GDR-wide internal council of astronomy research, which in my opinion was far too often concerned with the planning of a Large Observatory of Socialist Countries (OAO) full of super telescopes on a high mountain in Uzbekistan, instead of looking for the short distances to the existing instruments on Calar Alto or Tenerife. This was reminiscent of Lauter's disputes about the weighting of COSPAR and INTERKOSMOS. Because of this comment, Treder as the chairman of the council abruptly ended my membership there. It had already been

Fig. 9.3 Left to right: Manfred Schüssler; Michael Knölker. Potsdam-Babelsberg

communicated in April 1989, however, that the GDR contribution to the future super observatory had to be based on the profitability of Carl-Zeiss Jena. The most that could be contributed was the production of 2-m-mirror telescopes financed by all countries. The head of the astronomy section at Carl Zeiss, Köhler, had the courage to realise that we do not yet have any expertise in important technologies such as computer control, active optics and CCD, which are indispensable today. Only a few months later, the dreamed (15…25)-m mirror and the 2-m solar telescope in the Altai Mountains had suddenly disappeared from the earth's surface and the East German astronomers observed at Calar Alto and Tenerife.

On 9 October, a Dresden diurnal newspaper reported that after a concert, a visitor stood up and shouted to the audience that the seriousness of the situation had not been recognised in Berlin up to now and "SED membership can no longer be a precondition for top positions". The excited gentleman was Baron Manfred von Ardenne, Dresden's most famous scientist. At the end of the year, the first flags in the green-white colours of Saxony appeared on the houses, and their number grew rapidly until their meaning became obvious: it was the residents' subtle expression of "Adieu, GDR", more subtle than in June 1953 when lorries full of workers with slogans rattled across the central square.

On 18 October Honecker fell from power to homelessness after almost two decades of autocratic regime and Egon Krenz—who had just returned from Beijing— immediately took over. *Mr. Tschäpe has read the Krenz-speech, it really isn't that bad. Krenz doesn't go to the ideological foundations, but in this situation that can't be done* reads a wiretap transcript from 20 October. *Mr. R. is curious to see what will happen in Leipzig on Monday, Mr. T. predicts it will stay as last Monday.* If the illegal listeners of our telephone talk believed Rudolf they would have been horrified by the 300,000 demonstrators in Leipzig[10] only a few days later.

On 4 November, an open letter asked Ruben how applications for membership of the IAU were handled in the past. Ruben with his typical naivety answered that the institute leaders discussed these applications and submitted a list of candidates to the General Secretary for approval. "Those who fulfilled the necessary professional requirements and the conditions for travelling were included in the list." He ignored the fact that he himself had arbitrarily created or denied the conditions for travelling under total ignorance of scientific issues. The internal status of the employees was formally determined by the institute director but the so-called advisory group made the decisions. In

---

[10] Around 10,000 demonstrators on 2 October 1989, 130,000 on 9 October, 300,000 on 23 October and 500,000 on 6 November.

ZIAP it consisted exclusively of the political nomenklatura including the IM "Astronom", "Walter" and "Kosmos" in 1984. General Secretary Grote had already offered the institute directors this idea in July 1976: *I would like to draw your attention to the fact that applications for foreign visits can only be made for scientists who are confirmed as travellers.* Folks, it certainly meant, soon there will no longer be any duty for success in your science!

The presidium of the Academy commented on research in an overlong statement not until end of October 1989. The basic failures in the science management only got the almost trivial ideas that international exchange in science may deserve all encouragements and that administrative obstacles "such as the travelling status must be removed". Much too late, in a few days all gates will be open and many scientists will start their real live without having to follow the self-serving rules of overburdened science commissars. On 10 November 1989, despite celebrating the opening of the Berlin Wall, almost a thousand younger employees of the AdW gathered close to the Academy headquarter for a protest rally, at which President Scheler also spoke but under protest. He claimed that the AdW had already submitted proposals for a new strategy for academic research to the government. He reported that at the latest meeting of the ZK of the SED, of which he himself was a member, the separation of politics and science had been discussed but he didn't find own words for this basic GDR defect. Was this the top director of the largest scientific organisation of the country or did he speak as a dependent apparatchik of the state party; the massive personnel dilemma of socialistic science functionary had become obvious for all listeners. Perhaps in response to the loud protests recorded by the MfS, on 7 December he—a party member since 1945—asked for confidence in himself, his vice presidents and the GS Grote by the plenum of Academy members. A handful of Academicians had demanded this. Scheler received 77 out of 86 votes; Grote, who was responsible for the two-class personnel policy, and the vice president for social sciences were victimised. Instead, the philosopher Hörz reached the upper level as a new vice president. He will later be voted out as director of the Institute of Philosophy. Because they had imagined the needed transformation more strictly, 21 astronomers in Babelsberg subsequently declared in an open letter that the mandate given to the new presidium on 7 December 1989 does not correspond "to a mandate from the staff of the institutes and cannot replace it".

Speakers on the AdW protest meeting also demanded that the separation of research and politics must imply the complete absence of SED organisations in all academy institutions. The council of institute representatives, which had formed from the reform movement in Berlin, demanded on 19 December that no political party should be allowed to work within the

AdW. The Presidium did not comply; an instruction from the President to dissolve the central SED organisation at the Academy or even the local institute groups never appeared. Until the summer of 1990, the council of institute representatives had no office of its own, no photocopier and only a few private telephone lines. In May 1990 the activists were given a list of the institutes and facilities with postal addresses and the related telephone numbers. Under these conditions, only a few Berlin institutes were able to take notice the existence and the minor public impact of this new academic player.

On 7 December, the plenary session of the academy members had instructed the modified Presidium to continue business until the AdW was reorganised. The election of a new President was scheduled for 26 April 1990. The so-called Round Table of the AdW set up to prepare for the election was chaired by the colourful H. Klenner, an experienced theorist of law. His main task was to secure the privileges of the Academicians. As late as 31 October, he secretly declared as IM "Klee" that the MfS must find the conditions under which the transformation can take place in a direction that "corresponds to our Marxist-Leninist conception".[11] Strangely enough the Round Table followed his ideas that neither non-Academicians nor any external scientists should be considered as candidates for the new presidium and, even more strange, that the Round Table should work in secret to exclude external influences. The latter also implied that the ZIAP's proposal to nominate the founding member of Neues Forum, Jens Reich, Professor of Biomathematics at the Central Institute of Molecular Biology, as president candidate had become obsolete. The public had no opportunity to comment on or even to favour such proposals. The naive and inexperienced reformers had given away its two greatest trump cards—openness and publicity—without any need.

With the molecular biologist Heinz Bielka, a presentable candidate, was available who had not been a member of the SED for a long time; no one else from the whole plenum had run forward. This in no way corresponded to Klenner's Marxist-Leninist conception, and the election day on 26 April was thus cancelled by him due to a "lack of choice"—which never before in socialistic societies was a reason to cancel an election. What's more, he also justified the proposal of the Round Table to adopt the present draft of the statute without further revision, despite the many opposing opinions received. The new Statute of the Academy of Sciences of the GDR, which was to come into force on 1 June 1990, included the "Community of academy members" and the "Community of institutes" on an equal footing. The academy members continued to consider themselves authorised to decide alone on the appointment of

---

[11] Stark (1997).

professors. The AdW also still insisted on the right to award doctorates with the Soviet levels of doctor of a branch of science and doctor of sciences, which was strictly incompatible with the Western academic system. The postulation of the—completely needless—unity of the Academicians and the research institutes pervades all official papers of the academy management from this period. In August 1990, a short declaration by a handful Academicians including Bielka and Hörz still whisked the "Gelehrtensozietät", the "Gelehrtengesellschaft" and "Our academy". At the same time, the President elect warned urgently of disaster if the research institutes were to be assigned directly to the future federal states and their governments. This the more as the two science ministers emphasised that the institutes are to be transferred to the responsibility of the newly formed federal states in accordance with the constitution and the practice in West Germany. None of those alarmists has ever asked whether the future eastern institutes and institutions might not also work without any plenum of Academicians who like to appoint professors. Nor did anyone ask the rectors or presidents of the universities how they felt about sharing the right to award doctorates with the AdW in East Germany forever. The old elite finally argued as if they had fallen out of the world.

On the rescheduled election day on 17 May 1990, Bielka suddenly had four opponents: Herrmann (history), Klinkmann (medicine), Lohs (chemistry) and Peschel (mathematics). Herrmann and Lohs had been party members for decades, Klinkmann even a functionary as a member of the SED central headquarters in Rostock. The election assembly consisted of 100 voters each from the plenum, the new research councils of the institutes and the employees. Klinkmann with his own vote received 151 votes in the first ballot, just the majority required for election, Bielka 62 votes and Lohs 61 votes. Klinkmann was later dismissed from his university in Rostock due to lack of personnel integrity. The state government followed this decision, but any detail, apart from the accusation of being too close to the SED state, remained unknown to these days.

The second appointment of the day for the position of Vice President for institutes and facilities was decided in a similar way. S. Nowak was elected, the former head of the Chemistry Department, a party member since 1948, who also had a hidden career over decades. The determination of the GDR nomenklatura to hold their positions was obvious; on this day the apparatus proved itself as incapable of real reforms. There were dozens of similar examples of the astonishing endeavours of the friends and participants of the old regime also to dominate the new academic system—if necessary by democratic means.[12]

---

[12] Meinhold (2014).

The new government in East-Berlin elected in March had problems with the persistency of the AdW. Also the new academy leadership had the ambition to establish the academy members and the institutes together as a fourth pillar in the all-German science system. Only on 27 June, the resolution of the government "on the further activities of the Academy of Sciences" dismissed Scheler, and Klinkmann was appointed as the new President. The final reaction of the politics came exactly one week later, on 3 July 1990, at the legendary fireside chat between the ministers Riesenhuber and Terpe and the Presidents of the scientific organisations of both Germanies, including Klinkmann and Nowak. It was agreed that there should be a homogeneous research landscape for the whole Germany, without a special construction for East-Berlin. The Wissenschaftsrat was requested to organize an evaluation of the research capacities of the GDR and to make proposals for their reorientation. The manifest stops with the assurance that „the own approaches to evaluation visible in many AdW institutes are welcomed and should be utilised and incorporated".

Article 38 of the Unification Contract (Einigungsvertrag) says that the Academy of Sciences of the GDR will be separated from the research institutes and that an assessment of publicly financed institutions by the German Science and Humanities Council will be completed by 31 December 1991. The contracts of employees will continue to exist until 31 December as limited contracts with the federal states. With the unification of both Germanies on 3 October 1990, the AdW ceased, therefore, to exist as a research organisation. Its liquidation ended the takeover of all the historical institutes by SED functionaries that had begun with the academy reform at the end of the 1960's; the following decades of stagnation had come to an end. With "The Academy President no longer exists. All our worries, all our hopes are at a new beginning", the president for a few weeks formulated his farewell to the employees of the former Academy of Sciences of GDR with their limited contracts for still 15 months.

German Democratic Sternwarte

The tanks had stayed in the barracks, the fear was gone and not too much was left in the hands of the former nomenklatura. The employees were even given their own, previously top-secret personal files for safekeeping, albeit freshly laundered; the files contained little more than a handwritten CV. It was now no longer possible to avoid discussions about the scientific secretary Helbig with his exceptional income. He was briefly employed by ZIPE as a security officer in the 1970s and moved to the team of the new and neutral institute director K.-H. Schmidt at the beginning of 1984. Some people realised immediately Helbig's legend: he was a full-time FIM of the MfS who already guided the astronomers "Astronom", "Karl", "Schmall" and "Walter" among others. His precursor had made a commitment to the MfS to carry out six secret meetings a week and Helbig had taken over this network.[13] With six interviews per week and a monthly rhythm of contacts, a large number of informants from several academy institutes must have been managed. The MfS had sent one of its most hard-working agents to the astronomical institute. How bizarre it must have looked when "Jochen Gränz" was insisting on the opinions of his professorial informants? Ruben explained the world of Professor Krause to him on 25 January 1982 in a conspiratorial flat: He *focuses on his own work and maintains his reputation as a scientist, which many colleagues at this level lose because of their administrative duties. K. often turns to me asking how to tackle a problem.* Unfortunately, it must remain open forever which problem Krause wanted to talk over with "Astronom". It is not known whether the personal code names were used during the conspirative meetings and whether they left the locations with dark glasses.

In January 1990, an open letter appeared in the Sternwarte. The argument that it was incompatible with the spirit of the democratic transformation that Helbig remains involved in the top management of the institute met with broad applause. Almost all non-party scientists in Babelsberg signed the letter, with only a few names missing. Helbig was hardly ever seen in ZIAP after that. At the beginning of February, an ad appeared in a Potsdam newspaper: "International capital providers offer involvement in the conversion and extension of real estates. Contacts via Dipl.-Ing. J. Helbig."

Much later, at the end of 1993, there will be another open letter addressed to the IM "Wolfgang Bley". Stefan Gottlöber wrote to "all colleagues of the AIP, to all colleagues of the former ZIAP: Some time ago I took a look at my documents at the Gauck office[14] and realised that you had also reported extensively about me. When we discussed contacts with the MfS in several meetings at our old institute, you firmly asserted that you only had contacts with the MfS in the context of your cooperation with the army. That is obviously not true." As early as 30 January 1991, the Internal Scientific Council (ISC) of the Institute, in the person of its speaker, had requested that scientists who express a wish to work on one of the two lead projects then they had to declare that on oath that I have never worked as an official or unofficial employee of

---

[13] Buthmann (2020), p. 1100.

[14] This organisation managed the written protocolls of the former MfS.

the former Ministry for State Security. I will agree to a review of this statement." This procedure was considered so self-evident at that time that even the three former informants "Dr. Mann", "Walter" and "Wolfgang Bley" in the second ISC also voted that in consequence approved secret informants would have no future at the new projects. As the code name "Wolfgang Bley" also appeared in the files of other employees, his old declarations proved to be incorrect; he left the institute soon afterwards.

The tragedy of socialism is the separation of intellect and power, Heiner Müller wrote ironically after the colossal free demonstration in Berlin on 4 November 1989. The disputes around the turn of the year 1989/1990 about a new institute statute belonged to this headline. The most urgent tasks were to rediscover the scientific output as the people's main performance and to reorganise the institute in a modern-science oriented way. Both could only be done on a fair and transparent basis through a suitable election system, because the mere appointment of a new president or by parts of the staff—as the old GS had envisaged—would hardly have been effective. The development in the presidium demonstrated the risks. Even the central SED commission mentioned at the protest meeting on 10 November were more interested in its own fate than that of science, one had thus to help oneself. On 4 December, an internal ad hoc group presented a first draft of a charter for an ISC, the final version of which was adopted by a staff meeting in January 1990. The council was to consist of nine members, one of whom was to be elected in each of the Tautenburg and Sonneberg observatories. It declared itself to be the highest scientific and science-management authority of the institute, which also made future guidelines from the Academy President subject to its approval. By the draft version, the ISC also appoints the director; in the confirmed institute statute it only advises him. The ISC was to evaluate the concepts of the project groups and determine the distribution of the available funds. The "projects", which were originally intended to be temporary, became the basic elements of the future structure. Initially, everyone was allowed to declare themselves as head of a defined scientific task but every researcher who is not pursuing an own project in this sense must join another project. If he is not accepted by a project leader, he must be assigned elsewhere. Only if several scientists want to work on a project, it can be confirmed by the ISC or not. This was the self-organizing way of ending the epoch of polit astronomy in the GDR. As a result, the evaluation commission of the Wissenschaftsrat will find at 5/6 December 1990 in Babelsberg, Telegraphenberg and Tremsdorf a restructured institute with ambitious ideas for present and future research.

The small lecture theatre in Babelsberg was now always packed at seminars and colloquia, even when a member of staff from the Nuclear Research Centre in Rossendorf near Dresden gave a talk on technical applications of magnetohydrodynamics. Gunter Gerbeth had completed his doctorate in Dresden the previous year with Krause and Lielausis (Riga) on the examination panel. He had spent months working in Grenoble and reported from the laboratory of the Institut de Mecanique that interested me most at that time. Could electrically conductive liquids such as mercury, gallium or sodium be magnetically influenced in experiments changing the symmetry properties of flow, possibly to the point of generating two-dimensional turbulence? I never forgot this talk so that—once own ideas about possible MHD experiments emerged much later from the studies about magnetic instabilities—it was immediately clear that the partners for their technical realisations were not too far away from Potsdam.

The ISC was to meet monthly. But already on 19 March 1990 there was an occasion for an extraordinary meeting. The long-standing director, K.-H. Schmidt, had asked the AdW president to dismiss him for health reasons and asked the ISC to discuss his succession (Fig. 9.4). At the end of 1989, as his own evaluation and preparation for new and more critical times, he had counted the publications of all his scientists in Potsdam/Babelsberg, Tautenburg and Sonneberg under the heading HOW GOOD ARE WE? He compiled the papers in 5-year blocks taken from the Astronomy and Astrophysics Abstracts from Heidelberg and compared them with corresponding results from German-speaking countries. This was a typical anticipation of him to the use of statistical methods to measure scientific

Fig. 9.4 Karl-Heinz Schmidt (1932–2005), chair of the Universitätssternwarte Jena 1978–1982, director of ZIAP 1982–1990. (Photo M. Schulz-Fieguth. Courtesy of M. Schulz-Fieguth)

performances of persons or institutes which is common practice today. The far more effective method of counting the citations of publications would also have been possible at that time, but with much greater effort. His results are not easy to read because almost all documents have been counted, hence only the tiny numbers are of particular interest. In the period 1984–1988, the 6 Section heads—with an average of 9 citations—were below the institute's average of 11; three of them had almost no written contribution to astronomy or astrophysics in the past 5 years. The construction of the entire institute proved to be almost inverse. In fact, political reliability played the main role for a career in the GDR, while professional success was of secondary importance. This was inherent in the system, not a coincidence. It could not go well in science Schmidt will resignating have realised. One can understand that the former Section heads—only one was without party membership and four of them had an IM status—only played a minor role in the upcoming structural regulations.

Already in June 1984 the "scientific secretary" reported to the MfS that there were *discrepancies between Schmidt and the deputy director, who unfortunately was also party secretary* with problems to hide his own ambitions. In consequence Schmidt joined the state party SED in 1986, and the deputy director lost his position as the influential secretary. As Schmidt later explained, his position had become hopeless against the power of the local SED party organisation. After the formation of the new federal state Brandenburg he wrote to the new elected Prime Minister *with great concern for the future of astronomy in East Germany*, stating that the coming structural decisions should *justly take into account the complicated political and technological circumstances for which I am morally responsible, too.*

The opportunity for a complete reorganisation of the ZIAP had arisen sooner than expected, but instead the ISC had proposed to appoint Liebscher as successor to the top position. An additional explanation was given that it is also planned to reform the entire institute administration. Proposals in this regard were to be submitted by the end of that year, but this did never happen due to the disappearance of the AdW. On the other hand, Rädler as speaker, explained in his report a year later that he was asked to submit several proposals for filling the position if possible. In the absence of other candidates who were prepared to take on this task, the ISC only nominated Prof. Liebscher. This left open the question of how intensively an alternative to the solely high-motivated candidate was sought. It is not easy to understand why none of the elected councillors declared their willingness to stand as a candidate; this reservation may be due to the tense political atmosphere exactly one day after the first free parliament election in GDR. With the exception of an area around

Berlin, which later became the state of Brandenburg, the former block party CDU was the strongest party everywhere. The result of this election implied the rapid liquidation of the GDR.

A few of the former party activists in Potsdam wanted to capitalise their reputation bonus in the West German community as quickly as possible and were counting on an immediate evaluation by directors of the Max Planck Society (MPG).[15] These may "form a judgement of the individual work as a basis for decisions, e.g., "for 'cutting old wood' at the institute". Obviously, the writer had quickly rushed into the new era. In contrast, the regulations of the Unification Contract (Einigungsvertrag) were comparatively mild. The reorganisation of the institutes had been planned as a multi-stage process, at the beginning of which own reform efforts would be made and at the end of which a founding commission would confirm them or not. As early as 12 October 1990, the Ministry of Research and Technology in Bonn reported that the staff of the ZIAP made an attempt "to restructure the institute and also to develop new research priorities", so that those research groups shall have a very good chance of continuing to exist.

In spring 1990 a letter arrived in Potsdam written by Wolfgang Wenzel in which he demanded clarification about why astronomers from Sonneberg were for decades never allowed to travel abroad for conferences or visitor programs (Fig. 9.5). Liebscher, the former officer in foreign affairs now replied as a member of the ISC and not as the new director. His answer was accepted by a majority of the council on 5 June but "Dr. Rüdiger declares that he does not agree with this letter".[16] How could I have followed this bundle of excuses, which claimed that "Wenzel was initially our first candidate. The institute's director [however] had the impression that the comparative assessment of Dr. Wenzel and Dr. Jackisch by the local political functionary had led to a blockade." If this had been true, it would not have been the German Science and Humanities Council that closed the Sternwarte Sonneberg the following year, but the local SED leadership there did it much earlier. It is not believable that some provincial party functionaries would dictate the powerful AdW—with its hundreds of full, corresponding and external members—who should lead the observatory behind the mountains; the strange idea that a completely unproductive early-retiree should serve as an opponent for Wenzel is part of the tragedy that was raised after Hoffmeister's death. The files reveal that the otherwise not understandable support for Jackisch came from Pätzold as the security chief of the Geo- and Cosmos Sciences Department in Potsdam. He

---

[15] J. Büchner addressed to the ISC, 12.7.1990.
[16] ISC of 4 April 1990 and 5 June 1990.

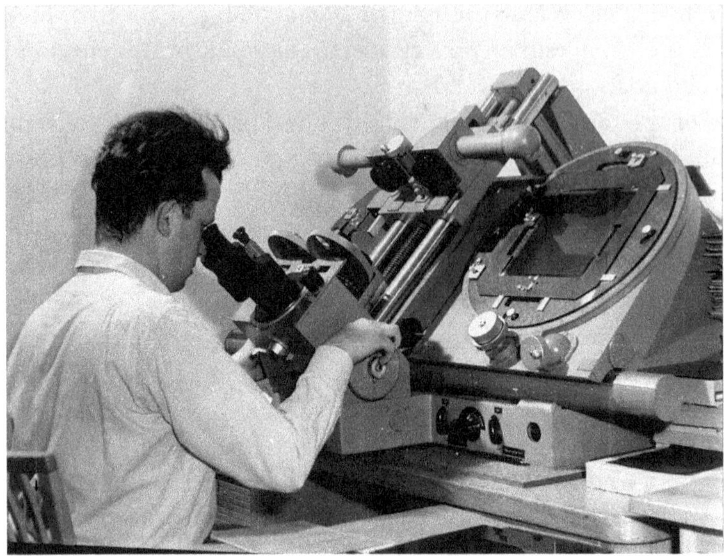

Fig. 9.5 Wolfgang Wenzel (1929–2021) on the blink comparator in Sonneberg. (Courtesy of H. Wenzel)

considered all employees of the Sternwarte Sonneberg except Jackisch as politically incorrect, whether SED member or not. "Astronom" complained to the FIM "Gerlach"—one of his leading MfS supervisors from the Department—that Pätzold stood for presumption, accusation and arrogance. He had written a dissertation at a very special Academy for State and Law (!) with the scary title "The transition from the antifascist-democratic state organs to socialistic power". This exactly was his programme for the future of the Department from nonpolitical research institutes to socialist-party structures. Wrong personnel decisions always shorten the lifetime of any sort of institution.

Wenzel refused from replying the narrative and wrote to the ISC on 18 June that he "was active to prevent the closing of the Sternwarte Sonneberg (as Profs Lauter and Treder wished), that our celestial cameras were destroyed after inspection (Prof Treder), that the bibliographic catalogue of variables was abandoned (Profs Ruben and Schöneich), that the telescopes were no longer maintained (administrative management) and that the Zeiss plate-measuring machine was withdrawn from Sonneberg (Prof Schmidt)". He asked whether it was not rather his problem with the official institute politics that had led to his rejection by the administration in Potsdam. Indeed, in August 1969 there had been an internal meeting in Sonneberg on the further existence of the Sternwarte. *Records of the discussion, which mainly centred on the closing of the observatory in Sonneberg, were unofficially written and sent out. Author and sender are not yet known*, reported IM "Astronom" to the MfS on 22 September 1969.

In July 1990, some council members held a discussion in Sonneberg about the matter, because Liebscher had given the justifications on behalf of the council. The secret list of all astronomers authorised to travel westwards, which was later discovered, indeed contains not a single name from Sonneberg. One of the top organisers of the personnel problems tries to shift the responsibility to subaltern functionaries far from Potsdam but the situation of the astronomy in Sonneberg was only the result of the bizarre personnel policy of the leaders of the ZIAP, the Department Geo- and Cosmos Sciences and the Academy. Wenzel as the undisputed head of the observatory had not been accepted as its leader, but even rather disavowed, with the active assistance of Sonneberg city officials and an isolated report-obsessed hidden informant. The reports by "Hagen"[17] were mostly directed against Wenzel. In 1976 the much-promising candidature of the latter as President of Commission 27 (Variable stars) of the IAU was prevented internally.[18] Jackisch had already finished his astronomical work in his late 40s producing a few petites for an astronomical school magazine. Letters of thanks for instructive lectures to regional SED functionaries he forwarded by express post to the headquarter in Babelsberg. On 3 June 1986, the formal head of the observatory, Marx, went to Sonneberg to meet Jackisch for talking about the director question. Wenzel's *constant intrigues* [were] *intolerable for the central institute, he had already discussed this with the director*, "Hagen" recorded and that Marx had explained him that only *his own directorship could lead to Wenzel's replacement as group leader*. That was wishful thinking, because *Jackisch is not suitable for this role in terms of his expertise and character*, wrote Ruben in a secret report.

The new director (from 1 January 1987) of the Sternwarte Sonneberg, W. Götz, will at the beginning of 1991 ban all his colleagues from the observatory on the day of the visit of the delegation of the WR.[19] Anyway, on 31 January 1991, a reduced working group[20] visited the observatory and talked with the majority of the scientific staff. In the written record of this meeting the scientific targets of the institute (variable stars in open star clusters; proper motions of variable stars) are briefly reported. However, *the suggestion that Sonneberg scientists should participate in academic teaching was met with*

---

[17] According to Wenzel's repeated applications to BStU: IM Hagen = G. Jackisch.

[18] Obituary Wenzel by P. Kroll, J. Greiner and R. Hudec, BAV Newsletter 2/2021, p. 120.

[19] Personal communication J. Greiner.

[20] Appenzeller, Haerendel (chairman), Jacobs.

rejection. The regular lecture on variable stars at the University in Jena stopped forever after Hoffmeister's death. "Let us do our research at last", Bräuer replied and Wenzel talked about the destructive attitude of Potsdam towards Sonneberg investment projects such as the 1-m mirror and the construction of a new plate archive.[21] Nevertheless, *in Sonneberg—in contrast to Tautenburg—a large part of the scientists were highly motivated*, the observatory's technical equipment, however, was seen as outdated, so that it may be not possible to carry out modern observing activities on this basis. *The KSO Tautenburg is more worthy of preservation than the Sonneberg observatory*,[22] was the conclusion of the commission anticipating the end of the Sternwarte in the mid-1990s. Wenzel continued to work and published until his 75th year. A late request by Marx to list his astronomical discoveries, he commented by hand in the top margin: It's over!

"To My Knowledge, Immediately"

On the morning of 1 November 1989, a lecture was scheduled in Babelsberg by MPG-director Joachim Trümper from Munich who wanted to provide information about his X-ray satellite ROSAT. After many years of preparation, ROSAT (ROentgenSATellit) was about to be launched at 1 June 1990, and its results will revolutionise the astrophysics just as much as the results of the gravitation wave astronomy have today. Trümper had written to the Universitätssternwarte Jena, the Institute for Cosmos Research (IKF) and the ZIAP some time ago asking for observation proposals for the X-ray telescope and had only received a single reply from the director in Babelsberg. ZIAP's involvement in space research before 1989/1990 was marginal: the INTERKOSMOS missions VEGA (Venus, Halley) and PHOBOS (Mars) were not successful. Later, at the end of 1990, several funding applications were submitted to the German Agency for Space Affairs (DARA) for the ASCHOT and SUV/T170 missions planned by the Soviets and generously funded by DARA for many years. The two evaluators at DARA from East Germany, Henning (Jena) and Rüdiger (Potsdam) had to solve the difficult problem to support the applications and to support only the qualified applicants. The missions, however, were never launched.

Trümper's appeal had only reached the directors of the institutes rather than the astronomers. The lecture room in the White House was overcrowded,

---

only the speaker was still missing. He arrived late because he had to listen the latest news from East-Berlin with the car radio. Some people in the audience laughed, others remained unmoved. Probably no one in the room knew that Trümper himself was a refugee. He had escaped the University of Halle as a student in the 1950s by train to Hamburg, secured only by a letter from his professor stating that a visit to a certain Hamburg laboratory was necessary for professional reasons and should be permitted.

In the evening, there was a semi-legal information event by Neues Forum in a church in Babelsberg, within walking distance of the observatory. There were more than thousand people, the speeches were broadcast outside over loudspeakers. Due to the many requests to speak, the talk time was limited to a few minutes. Trümper had found his way to the church, but he had to be satisfied with an external position. Finally with applause, a speaker announced the long-awaited, officially registered mass demonstration of Neues Forum for the coming Saturday. The speaker was Reinhard Meinel, who had just heard Trümper's lecture in the Sternwarte on the X-ray satellite. "The demand for far-reaching changes is widely supported in scientific circles", wrote Trümper after his return to the central administration of the Max Planck Society, in future every scientist will be able to travel, "provided the question of foreign currency has been solved. Nevertheless, in my opinion, one should be careful of the details in the MPG-Academy agreements, because we are not yet familiar with the practice."[23]

SED and GDR-state had not yet given up. A camp to isolate opponents have been located near Potsdam. In the safes of the over 200 MfS district offices sealed envelopes with collected names and addresses waited for use.[24] But the resistance grew explosively. Thousands attended the first election event of the Neues Forum Potsdam on Friday 3 November, the majority of whom were young parents between 30 and 40 (Fig. 9.6). The huge room could not cope with the rush and many had to stay outside. The demonstration the next day—5 months after the massacre in Beijing—will be the largest independent manifestation ever held in Potsdam, with many thousands of participants. From a balcony on a large city square the young priest Kwaschik explained the rule to sit down immediately if stones are thrown (Fig. 9.7). The demands on the demonstrators' posters were comparatively mild: Accept Neues Forum, Free Elections or The Country Needs New Men. Only one poster concerned the most political of all the problems, the Berlin Wall of cement and barbed wire: "Give me visa to Grandma Lisa". As the procession

---

[23] Letter Trümper, 9 Nov 1989. Archive of the Max Planck Society, III Dept., Rep. 139.
[24] Chr. Booß, Märkische Allgemeine Zeitung 4.10.2014.

Fig. 9.6 Erlöserkirche Potsdam November 1989: 2000 voters sent Meinel and Tschäpe to the provisional speaker group of Neues Forum. (Photo B. Blumrich. Courtesy of B. Blumrich)

Fig. 9.7 Neues Forum Potsdam: demonstration on 4 November 1989 with thousands of participants. (Photo Klaus D. Fahlbusch. Courtesy of K. D. Fahlbusch)

set off along a wide magistral behind the bit enigmatic slogan banner NEUES FORUM, NEUE HOFFNUNG, thousands from the side streets joined in and the procession grew every minute until it took up the entire long street. This first protest demonstration since many decades was the largest event ever organised by one of the new civic movements anywhere in the country. For many participants it only ended at the military barrier system of the Glienicke Bridge—which will only exist for another short week.

After its last official action—the cancellation of the military education in the schools—the entire government resigned on 7 November, followed the next day by the rest of the old SED leaders. Late on Thursday evening, 9 November 1989, the Berlin Wall opened as a direct result of the government's hopeless attempts to formulate a practicable travel law, the realities of freedom and socialism could not be united. At a press conference, a journalist had asked the spokesman Schabowski of the new SED leadership whether the travel law presented a few days earlier had not been a big mistake and Schabowski replied: "A permanent leaving can take place via all border crossing points of the GDR to the FRG or Berlin-West" and—searching for words—"to my knowledge immediately". Some people from Potsdam, including Tschäpe, had taken this word for word and had successful driven to West-Berlin that night by car via the Drewitz/Dreilinden border crossing point. Ten thousands followed the next day, after which a police visa valid for 30 days per half year was required—this was the rest of the order that was still accepted to remain. Never before the police had worked so swiftly and well-organised through 100-m-long double and triple rows of people waiting for the visa stamps in front of a sports hall. A day of anarchy: most of the offices were abandoned with open doors, theatre rehearsals were cancelled because the actors had run away and also the well-guarded Stubenrauchstraße 26 was empty. The same evening, the massive barriers on the Glienicke Bridge were moved aside. We could even return home at walking pace over the bridge from our first trip to West-Berlin between a lot of excited people with champagne bottles without drinking glasses. The two girls who hugged us and others on the bridge gave me a map of West-Berlin at my next birthday some weeks later. We could only listen to the cheering party with Kohl, Brandt and Momper in front of Schönebergs's town hall on the car radio because the flood of eastern Trabant cars through downtown West-Berlin. We were astonished by the whistling of the local political Alternatives, who had probably suffered an unexpected basic defeat.

The very next week, the small magnet group together with two cosmologists travelled the short loop from Stubenrauchstraße via the Drewitz border crossing to Wannsee in Mücket's impressive Russian limousine "Volga" to the legendary Hahn-Meitner-Institute for Nuclear Research (HMI). There was the small 10 MW reactor BER II, which fired the imagination of the anti-nuclear activists in West-Berlin. We hoped to find a large computer facility and perhaps be allowed to use it. At the entrance gate we asked for the computer department coming with our eastern identity cards. The head of the department was present and met us at the gate. Ulrich Nielsen—delighted that a horde of brave middle-aged theoretical physicists had been living on the other side of nearby Griebnitzsee for years—welcomed us with open arms (Fig. 9.8). After just a few hours, we indeed had institute ID cards, own email addresses and proper access codes for the VAX cluster. Elstner and the others installed their codes and got started, one of our biggest deficits had vanished overnight. Later, Elstner co-installed a new convex machine of the HMI at the end of 1990 with the experience he had gained during his visit to Wielebinski and Beck in Bonn in the spring.

The cosmology group of the astronomer fortress Stubenrauchstraße had used the observed "Ly-α Forest" in the spectrum of quasars to derive information about the temperature and density distribution of the hydrogen along the line of sight from Earth to the object of high redshift. Each cool cloud of hydrogen on the light path absorbs the hydrogen line emitted by the quasar at a slightly different frequency due to the expansion of the universe. Many clouds would lead to a whole forest of absorption lines on the violet side of the original line. For the cool clouds embedded in the hot intergalactic gas, a complex nonlinear temperature equation had to be solved numerically in order to arrive at the number and extent of the clouds compatible with the

Fig. 9.8   Left to right: Ulrich Nielsen; Detlef Elstner. (Photos private/R. Arlt)

observations. Just in time for the end of the ZIAP, a comprehensive paper about their works was published by the Akademie-Verlag. The focus was on the possible formation of structures in the very early universe through gravitation instabilities. That a close connection existed between this ansatz and the problems of structure formation, was one of their basing statements. Together with H.-J. Schmidt the various theories of gravity have been analysed to see whether their singularities provide Big Bang models or not.[25]

The research reactor in Wannsee soon played a role in Potsdam's city politics. At the beginning of 1990 in Potsdam, the later Federal Minister Fischer enthusiastically declared a green protest on the nuclear power station located on the border between Potsdam and Berlin. In response to my interjection that there is no power station, he replied that he had been told so. In fact, West-Berlin's Senator Schreyer together with some of Potsdam's left-wingers, was still involved in the campaign against the alleged massive danger of nuclear radiation in Babelsberg's villa districts. In a daily newspaper for Potsdam, with V. Müller we published the latest numbers: a nuclear power plant uses 3 tonnes of fissile material, in Wannsee it is 5 kg, the radioactive waste is equivalent to that of one of the ten large Berlin hospitals.

It was an inverse world for 6 weeks: until Christmas 1989 the border had only become permeable for GDR citizens but not for West Germans. Axel Brandenburg had heard about the fall of the Wall in Helsinki and quickly got himself a ticket for 19 November in order to do some dynamo theory in Potsdam. To his astonishment, he was stopped at the Glienicker bridge, unlike all the other passengers. He managed to call us for help from a payphone. Together with Meinel I picked him up on the west side of the bridge in my "Trabant", but we had to drive with him to the established Dreilinden border crossing. After paying the standard access fee of 25 DM per day, he was given a visa and we were all able to return to Potsdam. The watch tower at the gate to the Stubenrauchstraße was already empty. However, Axel, our first Western visitor, later complained that he only met us for lunch because of the permanent political events during the German fall.

Another construction site was the conference in Helsinki. Liebscher had already signalled in the spring that my participation—organiser or not—was impossible, Krause and Rädler would be travelling to the conference, just as to Prague in 1975. The decision was final, obviously I had a life sentence in a secret trial, the favourite punishment of all dictatorships. There was no hint for glasnost or perestroika, the country was in agony, who ever could, left. However, this time the MfS files—available a couple of years later—no longer

---

[25] Gottlöber et al. (1990a, 1990b).

contained evidence of external intervention in my case. The top-secret *Dictionary of political-operative work* stated that the *directors are responsible for selecting and checking travelling cadres.*[26] But that was not even half the truth as the approval of the Academy President for the institute's proposals was mandatory, who, however, had delegated the issue to a Department of Evaluation and Control of the Academy which consisted mainly of the security officers of the research departments directly connected to the MfS. Almost all travel matters went through this network, in which Stiller's bureau manager and security officer Pätzold was involved for all institutes of geo- and cosmos sciences. No wonder that the Sonneberg observatory, whose employees except for one remained suspicious of Pätzold to the end, never had a chance to get through this filter. Moreover, in his first life Pätzold was a SED commissioner for the history of the labour movement around the city of Brandenburg/Havel and school teacher for socialistic civic education which, of course, led to aversions to scientists with striking interests for either all-German culture or too many details of the real-existing GDR society. Younger nominees without political anomalies—mostly graduates of Soviet universities and party members—were later increasingly promoted after the experiences with the older cadres which usually remained invisible at international conferences. The quite visible SED secretary Büchner—also from Moscow—used his academy passport up to six times a year for activities in western states. Just the highest representative of the communist state party among the astronomers understood the isolation doctrine from 1967 for what it was: as a law only for the others.

The last chance to come to Helsinki would have been an order from the top SED secretary Jahn who had invited me to a discussion at the end of October after some of my public statements on science policy. I had hoped to impress him with stories about Academy and science because he had once studied in Jena. Before our date, however, Jahn was already so lost at his political position that a meeting would no longer have made sense, especially not with talks about science. He disappeared from the public shortly after the planned meeting. Nevertheless, one of his last sentences "A fight to the enemy, but we can't fight 10,000 enemies [in a city]" has gone down in historic memory.

We shouldn't miss the advent carol singing on 17 December 1989 in the Nikolaikirche, friends had told us in a whisper. In front of the church, on the way to the stairs, we met the current and last prime minister Modrow from the SED standing alone by his car, which looked nothing like a promo limousine. After a brief conversation in which we talked about joint efforts to

---

[26] Meinel (1994); See Appendix 8.

maintain the public order we went into the church. My wife looked for free seats for ourselves, greeting in many directions, while Modrow in his winter coat walked unnoticed sideways to his reserved seat in the front row. Suddenly there was applause, everybody stood up and greeted the entering Federal President as if he were their own. He was followed by a crowd of escorts, bodyguards, press and camera crews. In front, v. Weizsäcker turned to Modrow and greeted him with a handshake, again the applause, this time even from the choir. Later, the pastor preached that doors and gates have never been as open as they were at this time. After the singing hour, a convoy of black cars with blue lights on the roof left the old market place at a brisk pace, with my green minicar in tow. The destination was a press conference at Cecilienhof Palace to which I was able to enter with the support of Manfred Stolpe. Modrow told the journalists in detail about our personnel conversation in front of the church, that his government and new democratic forces such as Neues Forum had to work together to ensure the stability of the country until negotiations on a confederation of the two states or more could take place. The Federal President described his inner movement of driving across the Glienicke Bridge for the first time after 30 years. Weizsäcker also had a Potsdam episode in his life as an officer of the Wehrmacht Infantry Regiment 9. He said that we are one nation and that what belongs together will grow together, but it takes time to prevent it from growing bubbly together. Three different speeds had to be taken into account, i.e., the German dynamic, the dynamic in Europe and, thirdly, that the two alliance systems have to ensure our security, hence the three speeds must be harmonised. Stolpe would later say in one of his wilful interpretations that Modrow and v. Weizsäcker had announced the reunification of Germany that Sunday evening after the Christmas church service in Potsdam.

Tuominen in Helsinki had quickly found enough sponsors and scheduled IAU Colloquium No. 130 for the week after 17 July 1990, because the total solar eclipse, which was visible in Karelia, would take place the following Sunday morning. With "we ordered an eclipse for your pleasure", he will welcome the well-attended conference in July and will make jokes about the perfect predictability of solar eclipses in contrast to the accuracy of his own calculations in rotating turbulences. We had talked at length about the conference programme on the phone in December when, at the end of the conversation, he suddenly suggested that we should come to him on New Year's Eve, where Reinhard would be well placed to think about the German revolution. Overnight he organised accommodation and daily allowances, our tickets could still be paid with GDR money. The brand-new business passport, which was—suddenly—easy to acquire, contained a 5-year visa for travelling to all countries; I had marvelled at the stamp more than once. On the last day of the year, the plane took off from Berlin-Schönefeld for our first trip to the West as astronomers, for which the socialistic state had first come into its final crisis. The only thing that could have helped against the two-class policy in the Academy was the solidarity principle of the protestant church in GDR which had always organised its trips abroad under the motto ALL OR NONE—but who found one or two privileged colleagues from the ZIAP who would have been willing to do so? When Meinel received the poisoned offer to switch sides towards the upper class with travel permission, he refused, saying it would be others' turn first. At the conference dinner in Helsinki, I announced as a toast to "all revolutionaries of the East" that Reinhard Meinel had realised just in time that in our country the government would have had to resign first to allow the Potsdam dynamo youngsters to attend this conference, laughter and applause. One cannot imagine what happened if we weren't able to listen the spectacular talk by Doug Hall on differential rotation and long-term photometry of star spots because of the veto of a bundle of apparatchiks. In the Foreword the editors of the proceedings underline that "analogues of the solar cycle are common if not universal amongst stars of similar age and mass. These developments have generated a new area of astronomy, loosely termed the 'solar-stellar connection'."[27]

After landing in Helsinki-Vantaa, we joined Ilkka with his rental car and admired the red granite blocks on the way from the airport. We lived in the Academy's guest house, close to the gleaming white cathedral on Senate Square. In the evening, we drove to the Käpylä neighbourhood, where Ilkka met the other guests for the New Year's Eve party in his home, including Lauri Jetsu with his wife, Axel Brandenburg and Ilkka's girlfriend Ulla. The snow-covered garden was lit with candles and torches, the house was not large, practical and full of waxed natural wood, Ilkka will spend his last weeks here in 2011. At the dinner, we had freshly caught salmon (not farmed!), smoked just a touch, along with tiny pieces of reindeer meat. At midnight, I promised the sceptical host that I would make up for the one-sided balance of visits in the new decade and see him in Finland as often as possible.

Over the next few days, the conference programme—Kwing Chan from Hong Kong should open the presentations with his first simulations of rotating convection boxes—and the expedition in midsummer light to the 300 m high Koli-Hill in Karelia were planned in Ilkka's office in the Observatory (Fig. 9.9). Meinel and Brandenburg discussed the nonlinear dynamo problem[28] and Axel

---

[27]Tuominen et al. (1991).
[28]Meinel & Brandenburg (1990).

Fig. 9.9 Observatory of University Helsinki, the site of the reception evening at 16 July 1990 of IAU colloquium 130. A conference photo does not exist

demonstrated on the screen the new text editor T_EX, which had just appeared and was about to make its way into the scientific world. The code, however, seemed too cumbersome to me, so I spent some more time with the easier-to-understand commercial ChiWriter, which the Deutsche Forschungsgemeinschaft (DFG) had financed for me, until I later realised my mistake while working with Schmitt in Göttingen about the (non-)possibility of swing-excited galactic magnetic fields due to the spiral arms.

Despite all the euphoria, there was still some uncertainty. There would soon be elections to the ZIAP's Internal Scientific Council in Potsdam, and I still had to write down the theses that each candidate had to submit.[29] At the turn of the year, a pressing issue had arisen for the new political movement because the Modrow administration had actually only renamed the MfS. On 5 December, activists such as Meinel, Tschäpe and Kaminski had the local headquarters of the old and new state security forces in Potsdam opened by a former stiff socialistic prosecutor under police escort. One could count almost

---

[29] First ISC of ZIAP: Aurass, Domke (until April 1990), Fröhlich, Liebscher, Marx (Tautenburg), Notni, Rädler (speaker), Richter (Sonneberg), Rüdiger (from May 1990), Staude (deputy speaker), elected on 23 January 1990; the former 10-person political leadership was reduced to 2 persons only.

thousand windows on the houses, with one or two people behind each one—that was the order of magnitude. I was requested to take on the feared MfS prison, the so-called Lindenhotel, together with my wife and a few other supporters. A linden tree in front of Hell's Gate, which should carry a memorial sticker, writes the former prisoner and later theatre director Kolkiewicz: Who knows how many last glances were fixed on its leaves, on its crown, how often it saw the disappearance, the farewells of the arrested?

We shouted "Neues Forum, let us in" through an intercom. Just the month before, this demand would have had horrible consequences for us, for decades the entire side of the road had been closed to civilian pedestrians, but on this particular day we had even been awaited. The gate through which thousands had been absorbed opened and the horror-servants were waiting behind a barred cage, some of them trembling. A former prisoner, Dieter Drewitz, arrested as a student in 1966, describes a later visit to the cell house where he was isolated many years ago. But then "the smell came into my nose that was still completely authentic. And then I made up what I never did during my entire time in arrest: I cried, it literally burst out of me."[30]

Almost all of the 200,000 political prisoners in the 40 years of the socialistic republic went through such cells; the prisons held people who had merely wanted to live elsewhere and were systematically locked up together with fraudsters, rapists and murderers. Not a single scene of an artist movie, theatre play or novel had ever been set in this GDR milieu. To intimidate him, Rudolf Tschäpe was questioned behind these doors for a whole day on 2 October 1989 about his political activities and family relationships. The MfS recorder may have been creeped out, but Tschäpe was *convinced of the correctness of his actions and the aims and intentions of the NEUES FORUM. He offered to speak about this to cadres from the SED, the state apparatus and other interested participants.* A penalty was imposed with payment of about a month's salary because of a political appearance in a church outside Potsdam. It was talked out, however, of the justice by the lawyer Gregor Gysi. Bohley, Tschäpe, Meinel and other activists had given Gysi the mandate to represent their interests in the event of a case. Neues Forum did not take part in the riots of 7 October in Berlin, Potsdam and other cities because of the foreseeable danger of violent clashes. Probably this was the information that the investigators wanted to obtain from Tschäpe on 2 October.

---

[30] Schnell (2005).

Fig. 9.10 MfS prison in Potsdam at 6 December 1989. During GDR times, a total of 6000 prisoners were held here, mostly men, especially many after 1953, 1961 and again from 1983

There were no more political prisoners arrested, apart from a few car thieves who had obviously been prepared to be presented to us (Fig. 9.10).[31] Our small liberation committee additionally met four women in their shared cell in prison. They were excited, hardly realizing the unbelievable: free people inspect the jailhouse and control the conditions. But they didn't want to become free, they were afraid of the uncertainties of this sort of

---

[31] In the 1970s, the GDR had imported 10,000 VW Golf, which were sold to (privileged) buyers for 25,000 M, but resold by 100,000 M as used cars. The prisoners had lived from the huge margins between buying and selling which was not allowed in GDR.

revolution, they feared the changes, just like their guards and, in a much weaker sense, just like us. The prison commander had cleverly made me personally responsible for the security of his house which was completely wired with signalling cables. Because we reasonably had left the doors to the street open so that courageous people dared to enter the building. Hence, I become thus busy with security and had to sort out the risky cases in the growing crowd, so that the interviews had to be cancelled and postponed until the next days. By then, the lockers and socialistic investigators had all come back, eager to know details of their future, they had seen influential people in us. Only one of them wanted to pursue a civilian profession in the future, all the others wanted to remain in a prison. In a letter to the editor that appeared later in a daily Potsdam newspaper, one of the MfS employees thanked us for the de-escalating atmosphere during the storming of the Potsdam Bastille—which, however, didn't stop some of his colleagues from destroying the tyres and glass panes of our Trabant in one of the following nights.

That was the situation everywhere. Consequently, the East-Berlin Round Table also required the dissolution of the remaining security forces on 7 December. Modrow actually followed, allegedly to reduce the successor organisation to a service for constitution protection, and a smaller secret service. As nothing of these promises had happened, the suspicion grew that Modrow only wanted to save as many as possible of his former comrades from unemployment. The GDR constitution was itself a case of reorganisation, what was there to protect by such a large service? All the breakfast discussions in the Finnish Academy's guest house centred around the topic of ensuring of what had been achieved. The convictions that Meinel and his friends developed in these days became visible on 15 January during the storming of the last remaining MfS headquarter in Berlin-Lichtenberg.

The Unified Sky

We used the days in Helsinki to further develop the solar dynamo model, in which electric currents and magnetic fields and their interactions are to be calculated simultaneously. The unknown variable remained the angular velocity of rotation, which, if chosen too small, does not allow the dynamo to work and, if too large, leads to incorrect rotation profiles. We called this problem the Taylor-number puzzle and it was to be presented in the summer as a solution to the dynamo dilemma. We also discussed the possibility of doing image processing for stellar surfaces together in Potsdam in the future. The beautiful plan, which could not go too far wrong, was later lost due to the multitude of influences of external advisors and other challenges.

On 12 January 1990, immediately after our return to Potsdam, there was a political spectacle of real Tschäpe-dimensions. He had convinced the famous TV moderator Lea Rosh from the public West-Berlin television station SFB to move her popular Friday talkshow to Schloss Cecilienhof in Potsdam. Even before the Wall came down, Tschäpe had already taken the journalist to the most exotic places of Potsdam's cultural history. My part in preparing the event was to obtain the necessary permits. The director of the old Prussia palaces, Mückenberger, a former head of the state film company DEFA who had failed because of the cultural policy, threw up his hands and asked if the Nazis were coming now? I was able to convince him that we are harmless and that he should check the result on television. Later, the chief of the Cecilienhof memorial excluded any direct contact with the historical furniture in the conference room (Fig. 9.11). During the show, prominent western people like Otto Wolff von Amerongen and Egon Bahr met the new Potsdam jumper fraction such as Saskia Hüneke and Reinhard Meinel. At neighbouring tables, dressed in fine cloth, waited the designated new party leaders Böhme, Gysi and Schnur their appearances. At the after-show party, Gysi offered us as much space as we want in the former SED daily newspaper for the press

Fig. 9.11 Schloss Cecilienhof, Friday night TV show; left to right: Tschäpe, Rüdiger, Meinel, Schnell. (Photo B. Blumrich. Courtesy of B. Blumrich)

releases of Neues Forum. On the basis of a special agreement, sofar we filled 2–4 pages there every week, which was difficult enough for the few after-work amateur journalists. A few days earlier, Lea Rosh, who was highly interested in all events in East Germany, invited the couples Meinel, Rüdiger and Tschäpe to a dinner in her home with predominantly European political table talks under her motto "Europe, what else?"

In Potsdam, we found a much-promising invitation from Deinzer and Knölker to a German get-together of several days in Göttingen, quickly and generously funded by the DFG. All German working groups on dynamo theory were to meet at the university's guest house in Reinhausen. In the early March Deinzer welcomed the numerous participants: "With the opening of the inner-German border, it was possible for the first time to come together unhindered, as was often desired by all sides." Stylish reception with fireplace, wine cellar and mild disputes on our common passion for the invisible magnetic fields in the sky, the complete Potsdam team had contributed to the meeting with talks. From Bonn Beck and Marita Krause came, Schüssler from Freiburg, Camenzind from Heidelberg and Brandenburg from Helsinki. Already over Easter 1990 Deinzer arrived at the Stubenrauchstraße for a return visit, also to familiarise himself with the remnants of the old Potsdam cultural landscape. He had offered me the opportunity to come to Göttingen in the winter semester to fill a temporarily vacant professorship at the university observatory. At that time, telephone calls from East to West were only possible via public payphones, which the federal post had very quickly installed in large number on the West-Berlin side of the Glienicke Bridge and we had often to speak about arrangements until I indeed travelled to Göttingen in October 1990 after I had replaced my Trabant minicar by a more comfortable and extremely reliable auto brand.

When the Executive Board decided in favour of Berlin as the venue for the 1990 spring conference, following an invitation from the Institute of Astronomy and Astrophysics at the Technical University, no one could have guessed how fortunate this choice was, begins the report of the AG spring meeting on "Accretion and Winds" in West-Berlin at the end of March 1990. Hardly any of the Potsdam astronomers didn't sit in the large lecture hall in one of these days. "Even long-time supporters of the isolation of GDR astronomy appeared at the conference", Dorschner ironised the situation in his published report in Die Sterne. All had travelled from Tautenburg, Jena and Sonneberg and stayed overnight with friends, relatives or in cheap East-Berlin hotels. My overloaded Trabant chugged along the famous AVUS every day. At that time, there was still free and unlimited parking at many places downtown in Berlin.

Besides Kudritzki's Munich stellar wind group, we wanted to meet the young Willy Kley because of his interesting boundary layer models of accretion disks (Fig. 9.12). Already after our first contact Willy proceeded in the evening with us to Potsdam, spent the night on our sofa and worked in the Stubenrauchstraße the next morning. He installed the disks with $\mathrm{T_EX}$-files from Helsinki on one of our new computers and familiarised us with his easy-to-use diffusion approximation for radiation transport. Much later he moved from Munich to London and Santa Cruz, from where Meinel recruited him to the Jena University at the end of 1992 as a specialist in numerical relativity theory. Meinel had left the ZIAP in March 1991 to work in the newly established Max Planck research group for gravitation theory in Jena with G. Neugebauer and others. Willy completed his lecture qualification there and eventually won the highly popular chair of Computational Physics in Tübingen. I met him again at almost every one of his stations, including in our garden plot in Potsdam together with his daughter and his mother, always admiring his very personal mixture of humour, light-heartedness and enthusiasm for science.

In April 1990, I boarded a train at Wannsee station in the evening which reached Freiburg on schedule in the morning after a 12-h ride. M. Stix was waiting at the station and led me across an overcrowded car park to his car parked far away. The beautiful villa of the Kiepenheuer institute for solar physics was located on a hillside near the cathedral and didn't look like hard work at all. Its lecture room also breathed the sunny atmosphere of the south. Director Schröter sat in an armchair at the back and smoked. Again and again

Fig. 9.12 Willy Kley (1958–2021) and Oliver Gressel (Head of MHD and Turbulence in Potsdam since 2020)

I had missed the opportunity to ask Grotrian's master students Mattig and Schröter about their Potsdam time. Only a few days after my 65th birthday—which nobody in Freiburg knew about—I had started a lecture in this room with the statement that I would be standing here for the first time if nothing else had happened in Germany. Some people smiled; it was a politically coded way of expressing my personal, long-standing admiration for this solar institute and its sunny staff. In the GDR, people have been allowed to travel almost anywhere in the world on the day after their retirement at age of 65, with their own money.

The first visit in Freiburg was about galactic dynamos—Stix had already developed a first model in 1975—and oscillations in solar-type convection zones (Fig. 9.13). Since in the Stubenrauchstraße a functioning network connection with the Hahn-Meitner Institute now existed, but the new workstation computers did not, Stix applied for a fast computer for the Stubenrauchstraße 26, which was still so expensive at the time that the DFG required a particularly detailed explanation of the project. For a time, this computer became the central unit of the house. The project that was supported by the DFG was to determine the influence of the structure of the eddy viscosity tensor on the lifetime of the modes of the 5-min solar oscillation.

In the same month, due to the rapidly approaching currency unification, an elected committee in Potsdam had to deal with the financial legacy of

Fig. 9.13 Michael Stix and Willi Deinzer (1935–2025) in Prague 1975. (Courtesy of M. Stix)

former director Wempe who died childless[32] in 1980. He left his patrimony of almost 900,000 M to the ZIAP "with the wish that the legacy be used primarily to improve the living conditions (e.g., by building housing) of the staff of the Astrophysical Observatory on the Telegraphenberg in Potsdam". Also, the first AOP director Vogel had left cash to the staff, resulting in the Hermann Carl Vogel Foundation with initial assets of almost 17,000 M.

A consultation with a legal office had established the situation that (1) the deceased's wish to use the money in particular for the employees of the ZIAP on the Telegraphenberg represents a moral, but not legally enforceable obligation and (2) the heir is permitted to pay out money to employees as the estate originates from private assets. The beneficiaries were defined as those persons who were employed on the Telegraphenberg at the time the testament was opened and who were still at ZIAP in April 1990. These 53 people were each paid 5000 M as their share of the legacy; the remaining amount, which was still considerable, was to be made available for the promotion of astronomical research at the ZIAP. Unsuccessful attempts were made to exchange the remaining sum at a ratio of 1:1 for the currency unification via the new State Secretary Helmut Domke in the Foreign Office of the last GDR government. The awarding of the Johann-Wempe-Prize for outstanding scientific achievements from the interest on the remaining sum was, in our opinion, the most lasting honour imaginable for the former director. The Wempe caretaker committee assumed that the will of the testator had been honoured as far as possible, also in the numerical relations. At its meeting in May 1990, the ISC "expressed no objections to the activities initiated by the elected Wempe committee".

In June, we visited the astronomical Max-Planck-Institutes in Garching/Munich. Tschäpe used his weathered Volvo luxury car with worn tyres—the trunk full of petrol cans filled in Potsdam—from the legacy of the famous painter Niemeyer-Holstein, which made us mobile. With E. Meyer-Hofmeister the stability problem of turbulent accretion disks was discussed; a presentation of the Lambda effect for maintaining differential rotation with Henk Spruit remained without agreement as did the presentation of my first attempts to attack the lithium problem of the sun by constructing the diffusion tensor for rotating anisotropic turbulence.

During lunch of 16 June, Director Trümper offered us to see the first light of his ROSAT telescope (Fig. 9.14). Due to foreseeable problems with our GDR ID cards, we were personally chauffeured by him to Oberpfaffenhofen;

---

[32] Also childless: Vogel, Freundlich, v. Klüber, Grotrian and Treder.

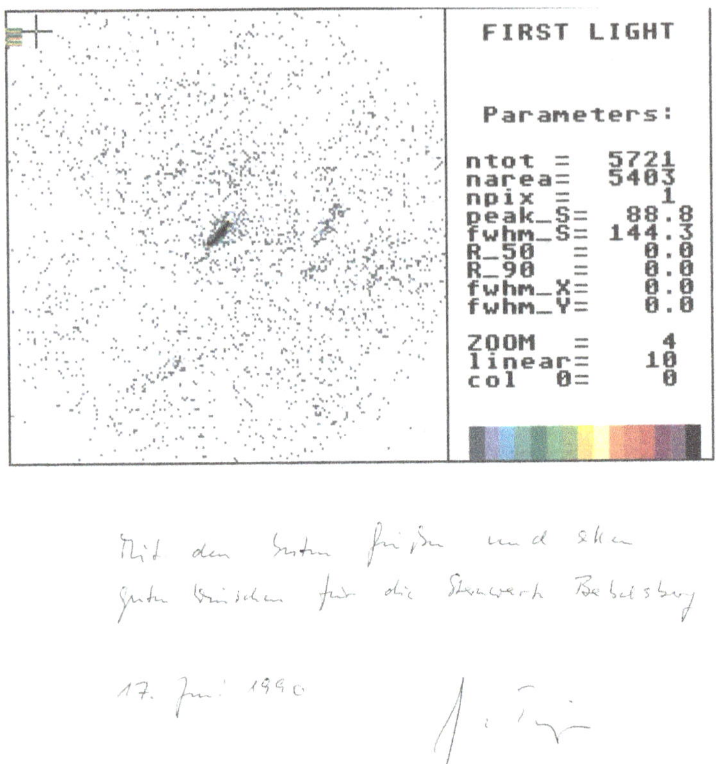

Fig. 9.14 ROSAT's first-light record from 1990, June 16, 23.45 MEZ, signatured by J. Trümper: With best regards and all good wishes for the Babelsberg Observatory, 17 June 1990. (Courtesy of J. Tümper)

a few words were enough to let us in. The control room was dark, full of monitors and busy people making adjustments. Suddenly, a large pile of dots appeared on the screens; the satellite up there had sent its first readable message and the screens were full of unknown X-ray sources. Nobody had ever seen this sky before us, we are living in a cosmos filled with hard radiation. One of the experts at the digital controls was Günther Hasinger who a few years later became director in Potsdam and soon after began to lead the Babelsberg institute. The next day, a Sunday, we rewarded ourselves with an excursion via Oberammergau and Garmisch-Partenkirchen to the Zugspitze mountain. Once on its top, we shared a portion of the famous white sausages because they always come in pairs. Tschäpe had the route for the return journey from Garching to Potsdam in his mind long ago: Nördlingen, Rothenburg, Nuremberg and Bayreuth. One couldn't have wished for a better guide through Bavaria which he never saw before. Since

Fig. 9.15   Jürgen Kurths. (Courtesy of J. Kurths)

this trip, I have also realised what a catastrophic giant caldera an asteroid of (only) 1 km in diameter will create.

At the beginning of July, Jürgen Kurths from the Tremsdorf radio observatory presented his project for astrophysical data analysis in the overcrowded seminar room in the White House (Fig. 9.15). He demonstrated the power of the modern nonlinear dynamics for data analyses as they occur everywhere in astrophysics.[33] Even deterministic nonlinear systems can only be finitely predictable, such as stock market prices, the rapid polarity reversals of the geomagnetic field, sunspot periods, high-resolution solar radio emissions, the eruption behaviour of the dwarf nova SS Cygni or possibly fractal density distributions in galaxies or galaxy clusters. New mathematical methods and novel measurement campaigns would have to be developed, focusing on bifurcation analyses, predictability studies and fractal behaviour. He also described the rapidly developing Doppler imaging of stellar surfaces as a standard method[34] of future astronomy and proposed an independent group for the subject in Babelsberg. Too many new things for an old institute; Kurths will later leave it to head the new interdisciplinary Max Planck working group Nonlinear Dynamics at the new-founded University of Potsdam.

With Alexander Hempelmann we planned to overcome the eternal AOP deficit of missing stellar calcium emission data (Fig. 9.16). We favoured an

---

[33] Kurths (1989).
[34] Hubrig & Kurths (1989).

Fig. 9.16 Left to right: Alexander Hempelmann; Jürgen Schmitt

existing, easily accessible 60-cm-mirror from Sonneberg, a newly constructed spectrograph and a highly specialised observation program. As early as October 1990, K. Stępień had already compiled a list of 50 late main sequence stars with known photometric periods and, if known, only low X-ray intensity. Differential rotation of the stars was to be determined as the main target of the project. Due to our theoretical results, fast and slowly rotating stars would have to be treated separately. In Sonneberg, a maximum of 100 observation nights would be available annually, an average of 2 nights per week. We hoped to be able to collect enough data for around 40 well-selected stars brighter than seventh magnitude in 7 years to derive the rotation curves. The project was called STELLA (STELLar Activity). Until the much later realisation on Tenerife—with other tasks and other operators[35]—we had unsuccessfully tried to locate a new telescope in Bulgaria, Israel or Australia (Fig. 9.17). Today, space-based telescopes such as GAIA provide calcium data from millions of stars.

In October 1991 in Tucson, we met Jürgen Schmitt from Munich, who was the opening speaker of the Cambridge workshop Cool Stars, Stellar Systems and the Sun with his talk "ROSAT observations of late type stars".

---

[35] Strassmeier et al. (2004).

Fig. 9.17 Construction (2005) of STELLA as a robotic telescope with two 120-cm-mirrors on Tenerife/Spain under the management of Klaus Strassmeier. (Courtesy of K. Strassmeier)

Hempelmann and me were suffering from jet lag, Schmitt apparently not at all; we got an idea of how high the bar would be set in the coming years. Schmitt presented X-ray data for the famous M-type star Proxima Centauri. Surprisingly, ROSAT had also registered X-rays from 12 individual stars of spectral type A, which should not exist there, but no signals from the brightest and best-known A-type star Vega. In Tucson, where we had set foot on American soil for the first time, we started to hire Schmitt for our STELLA telescope project (Fig. 9.16).

We as the two-men Potsdam delegation rented a car together but we had no idea how to use its automatic transmission and we also didn't know the speed limits. The clerk from the rental service explained unmoved that the speed limit is always written on the road. In the end we started and passed in the deep darkness hundreds or thousands of resting silhouettes of aircraft behind a kilometre-long fence. On every trip our car blew so much cold air between our legs that it soon became painful. Finally, by chance, we discovered how to switch the fan off. Ruben's permanent warning, "As a GDR citizen, one has to be careful abroad", proved to be correct in that, as a driver in Arizona one really should know what air conditioning is and how to switch it on or off.

By the late summer of 1990, the ISC of the ZIAP had compiled a review paper as a summary of astronomical research in this country up to the time of

Fig. 9.18   Left to right: Helmut Zimmermann (1926–2011); Hans-Erich Fröhlich. (Photo-Sammlung des Astrophysikalischen Instituts der Universität Jena/R. Arlt)

the political turning point.[36] Of the telescopes listed in this report, only the Tautenburg 2-m-mirror, the Einstein Tower and a 60-cm-mirror from Sonneberg will make it into the upcoming evaluation report by the German Wissenschaftsrat; the Sonneberg instrument, however, as to be transferred to the Observatory in Tautenburg. The memorandum was accompanied by a foreword by the chairman Pfau of the recently founded, short-lived Rat Ostdeutscher Sternwarten (ROS), which had to be hastily elected in order to get rid of the former political National Committee of Astronomy (NKA) which had previously represented the interests of the GDR as a socialistic state rather than its astronomers to the outside world. Helmut Zimmermann from Jena had proposed to create a democratically legitimised group through elections that could represent all East German astronomers (Fig. 9.18). He had a high personal and professional reputation since he had initiated in 1967 annual spring schools for all GDR astronomers, which were first held in a guest house of the Jena University and later in a hotel on the island Rügen. On 21 June 1990, he angrily reported that letters from Mr. Ruben and Mr. Gußmann "contained counter-proposals to leave the current National Committee unchanged". Zimmermann wrote that the entire National Committee had emerged from a despotism act, namely the appointment by the General Secretary of the AdW. It cannot speak on behalf of the majority

---

[36] Fröhlich & Marx (1990).

of astronomers in the GDR. He would personally resign if the old NKA continued as if nothing had happened in the meantime. The NKA in the GDR had originally been constituted in 1967 under the short-lived chairmanship of H. Lambrecht. The ISC of the ZIAP declared on 16 August 1990 that all the main initiatives for the dissolution of the NKA "were initiated by Prof. Zimmermann". In November 1990, the astronomical university institutions in Dresden and Jena were accepted as members of the Rat Deutscher Sternwarten. The future of the ZIAP had not yet been decided at the time of the meeting. [37]

---

[37] Mitteilungen der Astronomischen Gesellschaft 74, 9 (1991).

10

To Telegraphenberg and Back

Evacuation or Evangelisation?

Shortly after the German unification on 3 October 1990, the Federal minister for research demanded of the institutes of the former AdW that the Wissenschaftsrat will work as quickly as possible, but nevertheless it will be important that the institutes, on their own responsibility organize rationalisation leading to a modernisation of research work. A call for help, because all the (West) German commissions were not able to set up more than 70 renovated facilities from the old academy institutes. It would be best to use the time to reform themselves according to professional criteria in order to get favourable recommendations of the evaluation commissions. H.-E. Fröhlich warned that "evaluation" appears in the dictionary between evacuation and evangelisation. In August 1990 each institute had to submit a self-presentation to the Wissenschaftsrat, a detailed list of employees was due in November, and in December there would be a personal inspection of the ZIAP by a commission together with ZIPE, High Pressure Laboratory and the Meteorological Observatory. The second election of the ISC is dated to February 1991, the results of the evaluation are expected at the beginning of July, including the complete conversion of the payroll system to the western tables, and the founding commission must work in the fall as all contracts are due to expire on 31 December 1991. This timetable could only function if there were no incidents. The recommendations of the Wissenschaftsrat from July 1991 suggest the creation of 13,000 fulltime positions in nonuniversity research institutions, including working groups for the universities. Before 1989, the total number of staff at the AdW was 24,000. The number of 10,000 employees can already be found in first reform papers of the old academy administration from the end of 1989. Six institutes were completely closed.

In autumn 1990, the Max-Planck-Society announced its plan to set up temporary working groups at universities with institutional links to existing Max-Planck-Institutes and—with the prospect of a long-term commitment—to set up project groups as a preliminary stage to founding institutes or to establish Max-Planck- Institutes directly.[1] It exclusively concerned newly developing research areas, that lie outside or between established disciplines and are not yet ready for university research.[2] All of these options have been discussed for Potsdam and—with one exception—have also been realised for astrophysics.

During this time, the latterly isolated Krause applied to the International Astronomical Union for approval and financial support to hold a symposium entitled The Cosmic Dynamo with around 150 participants in Potsdam from 6 September 1992, shortly before the 100th anniversary of the death of Werner von Siemens, the inventor of the dynamo principle. After an international IAU workshop on Astrophotography 1987 in Jena with 36 foreign participants, Krause's project was only the second major international event organised by east German astronomers under the auspices of the IAU.

The application was successful and Krause will finally be able to top his lifelong attempts to bring together his dynamo colleagues from all over the world in Potsdam without any political restriction (see Fig. 10.1). The original application failed in September 1990 because the IAU did not want to support events dedicated to individuals. In the revised proposal, therefore, the name of Siemens no longer appeared in the title. It was the first conference dedicated solely to the self-excitation of large-scale cosmic magnetic fields, i.e. the dynamo problem. The response was enormous, with more than 100 foreign participants from 22 countries coming to Potsdam (Fig. 10.2). Schmitt from Göttingen followed by Kitchatinov from Irkutsk opened the congress with contributions on the solar dynamo. "Such young people are opening your great conference?" whispered science minister Enderlein to his seat neighbour Krause. The latter had just turned 65 and had just suffered a sudden hearing loss; he would soon be leaving the institute. His last publication—stating his private address and the MPG institute's name of his co-author Beck—surpassed many of the publications of the time in terms of

---

[1] MPG President, 26 Oct 1990.
[2] Presidential commission of the MPG, Sept 1990.

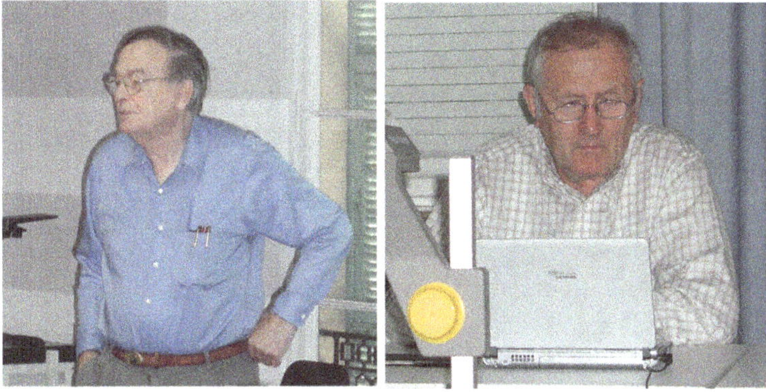

Fig. 10.1   Left to right: P. H. Roberts (1922–2022); R. Wielebinski. Potsdam 1992

Fig. 10.2   IAU Potsdam 1992 "The cosmic dynamo"; left to right: Rogashevskii, Léorat, Vishniak, Brandenburg, Rädler, Urbannik, Camenzind, Hempelmann, Tuominen, Stepinski, Meinel, Kitchatinov, Ferriere, Stępień, Kirchner, Rüdiger, Donner, Otmianowska-Mazur, Breitschwerdt, Khanna, Golla, Stix, Tschäpe, Kundt, Mestel, Roberts, Shukurov, Fröhlich, Ness, Krivodubski, Schultz, Beck, von Linden, Schüssler, Ferriz-Mas, Chiba, Duschl, Deinzer, Pohl, Ozaru, Lazarian, Rendtel, Schmitt, Sokoloff, Wielebinski, Krause (incompl.)

quality.[3] On 5 June 2007, the AOP dedicated an honorary colloquium to Krause on the occasion of his 80th birthday with contributions by Beck, Kurths and Stefani.[4]

In the summer 1990, all new projects on solar and stellar activity at ZIAP were placed under the common theme of solar-stellar-connection with magnetic field measurements of sun-like stars, actual time series analyses, Doppler imaging and star formation. The text on the new Lead project "Solar and stellar activity" postulated that activity "describes the status of stellar atmospheres which are either not static, not free of magnetic fields or not only heated from below", this was the philosophy for a modern AOP without magnetospheric physics and without the astrometry section of Ruben. Domke had left the

____

[3] Krause & Beck (1998).
[4] Appendix 10.

ZIAP to head the GDR Foreign Office in the de-Maizière administration as State Secretary. In this role, he also negotiated the 2 + 4 state contract between the four alliierts and the two Gernanies.

Once this concept had been given the final title "Cosmic Magnetic Fields, Solar and Stellar Activity", it became one of two Lead Projects in Potsdam, which was presented to the Wissenschaftsrat shortly afterwards with the title Astrophysical Observatory. Director and ISC of the ZIAP arrived at this self-description: "Following extensive discussions last year, the institute is focussing on four Lead Projects, which can also be divided territorially in the future. The project 'Extragalactic Astrophysics and Computer Centre' includes 28 scientists with PhDs, 19 other scientific staff and 27 technical staff. The 'Cosmic Magnetic Fields, Solar and Stellar Activity' project includes 35 PhD researchers, 16 other scientific staff and 21 technical staff." A total of (146/63) staff positions had been listed.[5] For the entire ZIAP, the ratio (250/160) was given, whereby the number 160 for scientists were not defined in detail.

The other two projects were formed by the observatories Tautenburg and Sonneberg, whereby in Sonneberg, in addition to the work with the plate collection for optical identifications of X-ray and gamma burst sources, the investigations of cataclysmic and symbiotic as well as short-period pulsation binaries were mentioned, no issue, however, was declared as central. With regard to the ISC, it is stated that based on the election results, the current institute director appointed by the President is also a member. "With regard to the management structure, it is planned that only the project managers will be subordinated to the director and the ISC."

At the end of November 1990, a revised ZIAP statute for the new ISC was formulated, which took account of the changes in the previous year. The secret election of the curators for the Lead projects was recommended which would appoint the members of the new ISC in a ratio of 4:4:1:1 with Tautenburg and Sonneberg. Only scientists who have publicly declared their affiliation to the projects were legitimated to vote. The speaker (Staude) of the ISC and its deputy (Müller) would also be elected by secret ballot by the council members thus defined. The new institute statute had taken account of growing centrifugal forces and their countereffects, and the composition of the ISC had changed accordingly. At the election meeting on 28 January 1991, the old (and new) speaker, Rädler, admitted that mistakes occurred during the preparation of the documents for the Wissenschaftsrat. When answering the question about politically motivated suppressions, the

---

[5] The second digit indicates the number of budget positions for scientists with a PhD. In the final version the term "Lead project" has been replaced by "Project complex".

difference between persons who were unable to accept a single invitation and colleagues who were never allowed to travel to conferences and meetings in Western countries was cancelled by the uncritical compilation of data, he said and didn't explain how this could happen. The final list of victims of the political circumstances even included six names of former privileged and very privileged persons, almost all party members. Already at its October meeting the ISC had stated that the answer to the question of the Wissenschaftsrat about restrictions on travelling in the past painted a false image of this time. The Babelsberg group already announced in April 1991 that the official organisation at the Babelsberg branch of the ZIAP would follow the project structure from now on: The project leaders determine the scientific profile of the group; they are in charge of the staff in disciplinary terms. The projects are subject to confirmation by the curatorship. The project leaders are accountable to the Chairman of the curatorship—clear instructions, still in old GDR language. Two days later, the ZIAP director finally replaced the outdated political institute structure with the new scientific projects at his last staff meeting. K.-H. Rädler was appointed director of the institute in May 1991 as Liebscher was dismissed from the institute by the Minister for Science on 18 May 1991. This decision was the result of the review of personal MfS files according to which Liebscher was active as the inofficial informer "Walter" between 1975 and 1984. The ISC shared the view that these facts are incompatible with the work as ZIAP director.[6] Liebscher had prepared a text (2 or 3 pages) and passed it around; "I thought there was a copy for everyone so that one could read it in detail. No, it was only meant to be passed around and briefly skimmed. There was a depressed atmosphere at the changeover."[7] K.-H. Rädler was immediately appointed as the new director by the ISC; hence the procedure of the failed draft institute statute of January 1990 was followed, because there was no longer any academy president existing. The final MfS report of 21 November 1975 on our cultural group which the officers called "Kontakt" already concerned on the recruitment of IM "Walter" for co-operation to *control over the effectiveness of the actions to destroy the group.* On 17 February 1982, "Jochen Gränz" had explained to the MfS how bourgeois scientists tick: "Walter" had reported to him that Roberts was the real channel for Krause's recognition in the West. *The two sought and found each other. They both get on very well professionally and personally. What always strikes me as odd is that K. is onesided in his promotion of Rüdiger while* **XXXX** *does*

---

[6] ISC of 21 May 1991.

[7] Personal communication J. Kurths.

not. This accusation meant that the independent scientists find western support and then they also support each other, unbelievable.

In the same month, there had been a conversation between Krause and his closest staff members in which he revealed his permanent MfS contacts, which were not—as had been hoped and accepted—only interested in information on trips abroad. This was never the case. We heard this report with frustration, but stated in a short Note that we do not assume that we have been harmed professionally by him as a person. The situation in the group was always science-friendly. He should continue his work on the preparations for the IAU meeting in Potsdam. On the other hand, we expressed our concerns about him holding public posts. Krause had subsequently left the executive board of the Astronomische Gesellschaft and a founding commission of the Brandenburg State University but continued the work on internal institute structures. The AOP, which still existed only virtually, had once again lost one of his leading figures. Meinel had extra arrived for this internal meeting from Jena where he has worked since April 1991 at the University.

After all, in the founding committee of the Potsdam University there were now no longer any physicists or mathematicians, and suddenly astrophysics was no longer included in the university's new curriculum. After realising this dangerous situation we compiled personnel statistics—on the advice of Max-Planck director Gregor Morfill—for all German university observatories with the clear result that the large astronomical centres in Germany are always correlated with large university institutes (Fig. 10.3). V. Müller and me asked the rector elect of the university, Mitzner, for an arrangement in this matter and presented our demonstrative data and a Memorandum signed by Rädler on the need of astrophysics chairs at the University of Potsdam. It explained that, according to "the ideas of the Wissenschaftsrat, the former astronomical world centre Potsdam, which is associated with the names of such researchers as Karl Schwarzschild, Albert Einstein, Paul Guthnick and Walter Grotrian should become the fourth centre of astrophysics alongside Bonn, Heidelberg and Munich". The new University of Potsdam should definitely secure the attraction of astrophysics studies, which results from the dynamics of the superfast developments in this field and from Potsdam's unique tradition. Finally, we modestly argued in favour of creating just two chairs for astronomy and astrophysics. However, the friendly and experienced physicist-chemist Mitzner only agreed to the creation of a single chair within the future Institute of Physics. This was a slow start, which would later develop into a flourishing institute centre with several professors, where the PhD students of the future large nonacademic Astrophysical Institute Potsdam could complete their doctorates once they had made it thus far.

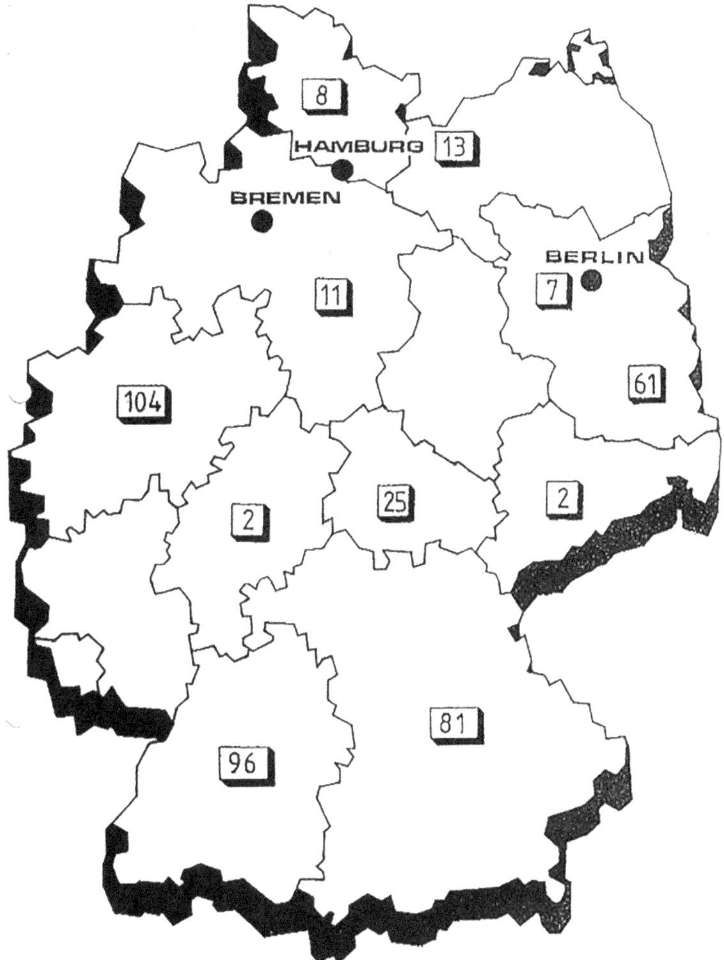

Fig. 10.3 Distribution of astronomer positions (without soft money) in the German states (end of 1990, before the evaluation). The numbers for East Germany only concern to scientists with PhD. Idea and realisation G. Morfill

The Grand Solution

In February 1991, the Lead project Cosmic Magnetic Fields, Solar and Stellar Activity took a further step with its newly adopted statute and demanded that the curators should campaign for the return of the buildings of the Astrophysical Observatory on Telegraphenberg. Two months earlier, a high-ranking advisory board of the MPE, chaired by G.A. Tammann (Basel) on the occasion of an early and voluntary evaluation of the ZIAP. In their report, the jurors expressed surprise that despite "isolation (for some scientists being complete), the great deficits of equipment, computers" and also the lack of students the research activities are on "a surprisingly high level". The commission found it also surprising that despite the special political situation of 1989, the productivity of the staff had not diminished. Despite all the limitations, it would only be possible to assess the performance of the work according to international standards. Only the two "magnetic" groups that emerged from the former Astrophysical Observatory, i.e., Optical solar observation and Dynamo/Accretion/Turbulence, were given the best possible ranking ("A") for the current status, while excellent future prospects ("A") were also awarded to the projects Extragalactic Astronomy, Image Processing, Astrometry, Plasma Physics, Solar Radio Astronomy and the Observatory Tautenburg. However, it was stated that some scientific groups "show a lack of leadership".[8] Obviously, Wempe's consistent focus on the exploration of the "magnetic universe" had paid off for the old AOP even after decades. The official evaluation of the ZIAP by the German Science Council will not come to a different conclusion either. The upgrading of the Einstein Tower for polarisation measurements carried out by v. Klüber from 1941 onwards and the development of dynamo theory on the Telegraphenberg initiated by Steenbeck have easily carried Potsdam astrophysics over all evaluation hurdles.

There had been misfire, too. Also in February, Lorenz and Krause drafted an alleged Concept of Reorganisation, according to which the Astrophysical Observatory on Telegraphenberg was to be revived as a regional institute supported by three to four Max Planck working groups—for which there were already formulated proposals—while the Babelsberg Extragalactic Astrophysics would be considered as a so-called Blaue Liste (BL) institute with a 50-50 federal and state funding. Apparently Lorenz, as the elected leader of the Babelsberg astronomers, had seen the testated quality of the Potsdam research not as a chance but as a danger for his groups. In 2009 the BL contained 86 research institutions outside universities of country-wide relevance, 36 of which were now in East Germany. It was originally supposed to be frozen in 1989. The science ministers from the new federal states, however, finally managed to increase the number of these institutions to around 80.[9] The concept of the MPG—who wanted to establish research groups at universities on a temporary basis and did not want to operate them to fill an otherwise half-empty regional institute—was completely misunderstood by the two authors. On 3 April 1991, the ISC took note of the MPG's position never to maintain

---

[8] Archive of the Max Planck Society, III Dept., Rep. 139.

[9] Personal communication H. Enderlein.

MPG groups in nonuniversity institutions and also not to participate in any mixed financing system.

The Rat Deutscher Sternwarten (RDS)—which was now also responsible for representing East German institutions—generally supported positions based on historical reasons: The traditional institutes of the astronomers in the new [eastern] federal states must be preserved in the future or restored. The latter applies in particular to the Astrophysical Observatory Potsdam, which should return to its traditional location on the Telegraphenberg. Apart from the future of the Sternwarte Sonneberg, the WR did not see it any differently. Its chairman Simon sent his final and fundamental report to the ZIAP director on 10 July 1991. Already the cover letter contained a summary: The Wissenschaftsrat recommends the dissolution of the ZIAP, i.e., the end of the 1968/1969 academy reform. Instead, "it recommends the establishment of three medium-sized institutes/observatories in Potsdam, Babelsberg and Tautenburg." A plasma astrophysics project group of the MPG should be located on Telegraphenberg and a centre for astrophysics should be established in Babelsberg; a Landessternwarte (Regional Observatory) should be founded in Tautenburg. Regarding the staff, it was stated that the employees identify with the institute. "However, due to the shortcomings in the technical equipment, the experimental groups are in some cases lagging behind comparable Western teams." On the other hand, the performance of the theory groups "meets international standards and is even among the world leaders in the field of cosmic magnetic fields and plasma physics. The performance of the dynamo theory group should be emphasised which must be maintained in any case."

For Potsdam, the concept included the formation of two separate institutions, a Project Group for Plasma Astrophysics on the Telegraphenberg with (45/20) positions and a Centre for Astrophysics in the Sternwarte Babelsberg with (45/19) positions for teaching at the newly founded Potsdam University and all Berlin universities. The project group was to be located on the Telegraphenberg using the historical buildings, such groups could be imagined as a preliminary form of a Max-Planck institute. The goal was to establish another astronomical centre in Germany alongside the three existing West German centres, including return of astronomy/astrophysics to the universities.[10] The Observatory Tautenburg was to continue to exist with a personal staff of (25/10) financed by the state of Thuringia.

According to this vote, the Sternwarte Sonneberg would no longer exist as scientific institution in future: The valuable plate collection and the archive

---

[10] Deutscher Wissenschaftsrat: Statements on the non-university research institutions of the Academy of Sciences in the former GDR in geo- and cosmos sciences (1992).

should be preserved and remain accessible for research by transferring them to Tautenburg. The entire northern sky had been monitored in Sonneberg since 1928 and around a quarter of all known variable stars had been discovered there. The archive contained at that time approximately 225,000 photo plates. The climatic conditions at the site (only 100 clear nights per year) and the outdated instrumentation did not appear sufficient to the Wissenschaftsrat of the continuation of these observations.[11] The climatic conditions in Sonneberg, however, are not different from those in Tautenburg. The Wissenschaftsrat was simply unable to propose that the small state of Thuringia may operate two separate regional observatories and that the decommissioning of a 2-m-mirror near the University of Jena was not really an option. The decision decades ago to locate Kienle's 2-m-mirror neither at Potsdam nor at Sonneberg had a high price in an unforeseeable way. If Hoffmeister had powerful brought this telescope into his influence, he would have secured the existence of the Sternwarte Sonneberg forever. The observatory, however, was closed as an institute financed by science ministries on 31 December 1994. The author's proposal to the prime minster of Thuringia to convert the Sternwarte into a Cool Star Research Centre with a budget of 5 Mio DM per year while retaining sky monitoring together with a university (Ilmenau, Erlangen or Würzburg) did not convince him simply because H. Ruder as the chairman of the AG impressed him with a another proposal of a budget of only 0.5 Mio DM per year.

In its soon formulated statement, the ZIAP's council spoke of adequate, albeit minimal, staffing for the Potsdam area and particularly welcomed the suggestion to offer over 100,000 students in the region a significantly broader academic education in astronomy/astrophysics. Commenting on the renaming of Cosmic Magnetic Fields to Plasma astrophysics, it was stated that plasma diagnostics should be understood to describe work on modern diagnostics of magnetic fields and activity phenomena on solar-type and very young stars. The plasma physics of the terrestrial magnetosphere was not mentioned further, which probably did not correspond to the intentions of the author of the evaluation text, G. Haerendel. The MPE then founded its own research group with Büchner (Potsdam) und Sauer (Berlin) for the physics of near-Earth space under Haerendel's direction in Berlin-Adlershof.

The recommendations of the Wissenschaftsrat also contained a few breaking points. According to its vote, the traditional observation-orientated stellar spectroscopy following Grotrian should be located in Babelsberg in the future. On the other hand, there was hardly any teaching experience at the Sternwarte,

---

[11] Elsässer (1995).

which was proposed to serve as a joint observatory for four universities, three of them in the state of Berlin. The scepticism of the Brandenburg government regarding the upcoming negotiations on the two-state institution was even more serious, also because the Potsdam University was only just being established and the reputation of the Senate of Berlin as a difficult partner was omnipresent. The Sternwarte Babelsberg would also have been disproportionately large for a university institute or a Landessternwarte. On the other hand, it became known that all Max Planck project groups would generally only be set up for 5 years and, according to frequent warning phone calls from Munich, it was not at all clear whether the MPG was even interested in the permanent operation of such a large facility in the historical buildings on the Telegraphenberg.

Brandenburg's Science Minister Enderlein had sought a second opinion in due time. He had invited Professor Trümper to his ministerial house in Potsdam on 17 April 1991, still before the report of the German Science and Humanities Council was published, in order to hear Trümper's idea on the future of the ZIAP in Brandenburg. In a letter to Enderlein dated 14 May, Trümper explained his opinion that the area of the former GDR should have a large astrophysics institute which "radiates to its surroundings and fertilises the teaching of the universities located there". He considered the concept presented by the institute which envisages two future directions to be very good. "It fits very well into the entire German research landscape." Trümper had only spoken here of *one* single institute with two branches. This formulation will reflect Enderlein's alternative intentions after the recommendations of the Wissenschaftsrat became known; in any case, his ministry initially put further plans for astrophysics in Potsdam on ice.

Federal Minister Riesenhuber who may have heard in Bonn about this development, insisted on the one-to-one implementation of the ideas of the Wissenschaftsrat, probably mainly to prevent the formation of new Blaue Liste institutes, which would have been half of his budget. In August, he landed by helicopter on the Telegraphenberg to explain his position to all those involved including representatives of the Brandenburg Ministry of Science led by Director Kleinhans. After visiting the Einstein Tower, he explained that not a single exception of the proposals for East Germany could be accepted, especially not in the case of astronomers, who are at the top of the alphabet. After the event, however, the members of the ISC were called together by Mrs. Kleinhans and asked with a recognisable tendency whether the foundation of a unified Blaue Liste institute based in Babelsberg would not be the best solution for all sides. Approving statements of some astronomers succeeded the sceptical ones. At the end, the Astrophysical Observatory on the Telegraphenberg would have fallen after all, despite repeated appeals from its curatorship, the RDS, Wissenschaftsrat and even the impressive helicopter minister from Bonn.

Only on 1 November 1991, director and council of the ZIAP sent out a statement on the current situation of the institute as a request for help to the public. Two months before the final dissolution of the academy institutes, the future of the astronomers was more uncertain than ever. "In the Potsdam area, no development in line with the recommendations of the Wissenschaftsrat had been happened. For some time now, the proposal to establish a Blaue Liste institute with 90 positions has been under discussion." This approach had already been reduced to (80/36). "We are now ready with detailed proposals for the establishment of a successor institution." *One* successor institution, not two—the wording definitively signalled the staff's agreement with Enderlein's Grand Solution.

Then everything happened very quickly. On 6 November the member of the federal parliament Emil Schnell received a note from Riesenhuber's ministry in which it was reported that an agreement had been reached with Brandenburg's government, Wissenschaftsrat, MPG and federal research ministry that a Blue List institute (financed in equal parts by the federal ministry and state Brandenburg) with 80 staff positions, 36 of which for scientists, will be founded. In addition, the MPG will form a working group with 10 positions. As chairman of the founding commission Prof. Trümper will be proposed. The Brandenburg ministry with Enderlein's alternative line had prevailed in all respects, at no more costal to Brandenburg compared to the original plans of the Wissenschaftsrat, i.e. a masterpiece by the local science administration. In September 1991 a presidential commission of the MPG headed by Profs Völk and Walther visited the old AOP. Despite a wide range of activities, the MPG had then not pursued the establishment of the plasma astrophysics project group on the Telegraphenberg. Instead, a Max-Planck working group at the University of Potsdam as well as three large and permanent Max-Planck-Institutes on the Potsdam-Golm university campus had been established.

In a letter dated 19 November, Trümper was appointed by Enderlein as chairman of the founding commission[12] of the Astrophysical Institute Potsdam (AIP) as an institute of the Blaue Liste. "By convening of a founding commission, the federal government and the state of Brandenburg want to create the

---

[12] Elsässer, Haerendel, Kudritzki, Mitzner (University of Potsdam), Trümper (Chair), Wielebinski, Schmutzer (University of Jena).

starting conditions for the future institute." Trümper opened the work of the commission in Potsdam on 4 December with a general staff meeting and explained that, according to the general regulations, exactly 90% of scientific positions—including all applicants with a university degree—will be awarded to former employees.[13]

The decisions of the commission were summarised by the designated director Rädler after their departure as a support for the necessary application letters. Of the 36 scientific posts, 30 went to the approved projects, while the remaining posts were needed for administration, workshops and computer technology. The confirmed research groups were MHD (Rädler), Dynamo and Accretion (Rüdiger), Optical Solar Physics (Staude), Corona physics (Krüger), Solar radio astronomy (Mann), Stellar physics (N.N.), Extragalactic I (Lorenz), Extragalactic II (Notni) and Cosmology (Müller). The scientist positions were divided 19:11 between Cosmic Magnetic Fields and Extragalactic Research, four positions were to be reserved for external applicants. The supporters of the planned Max-Planck project group on the Telegraphenberg had exchanged the spirit and premises of the old AOP for mostly permanent additional posts compared to the competing projects. The plasma aspect emphasised by the Wissenschaftsrat only appears with regard to the sun, but no longer in relation to the Earth's magnetosphere. A large and unified "Astrophysical Institute Potsdam" (AIP) starts its work in Babelsberg on 1 January 1992. The official founding ceremony will take place with ministers and the new mayor of Potsdam, Gramlich, on 1 October 1993 in the domed hall of the Great Refractor on the Telegraphenberg, with keynote talks by Professors Mattig and Trümper from Freiburg and Garching, resp (Figs. 10.4 and 10.5).

Also the working group about Nonlinear Dynamics of J. Kurths started at the University Potsdam, limited to a total of 8 years. Fourteen former ZIAP employees will move to various universities on 1 January 1992, financed for up to 5 years by the Scientist Integration Project (WIP). Only 16% of the 1500 employees of the former AdW supported by this program finally found a permanent position at some university. Only the Astrophysical Observatory Potsdam, the domain of outstanding researchers such as Vogel, Spörer, Hartmann, Schwarzschild, Hertzsprung, Freundlich, v. Klüber, Grotrian and Krause left empty handed. Its reputation finally disappeared from the memory of the astronomy community, almost at the same time as

---

[13] Meeting of the science ministers of the New States with the Federal Minister for Research and Technology Riesenhuber, Dresden 19 Sept 1991.

Fig. 10.4 Founding ceremony of the Astrophysical Institute Potsdam (AIP) with the founding director K.-H. Rädler (1935–2020) and Science Minister Enderlein. (Photo AIP)

Fig. 10.5 Ceremony of the Astrophysical Institute Potsdam (AIP) in the dome of the Great Refractor with J. Trümper as the chair of the founding committee. (Photo AIP)

the Sternwarte Sonneberg, the other location of the short-lived Institute of Stellar Physics. Krause will end his dynamo conference with the anecdote handed down by Roberts, in which Elsasser told Einstein that magnetic fields with a too simple geometry cannot be dynamo-excited according to the Cowling theorem. Einstein clarified that "if such simple solutions are impossible, then self-excited fluid dynamos cannot exist anywhere". He was wrong here, convinced that nature only follows the simplest solutions. After long unsuccessful attempts to prove the Cowling theorem in a mathematically rigorous way, our former colleague Jürgen Reichert had concluded that obviously there were important truenesses that cannot be proven. He died in a road accident in 1983, his manuscript on the all-dominant Cowling theorem has been lost.

11

Epilogue

As in our opening study "Astronomen, Akten und Affären" on the history of the Astrophysical Observatory Potsdamall quotations from the holdings of the Brandenburg State Main Archive, the Archive of the Berlin-Brandenburg Academy, the Archive of the Prussian Cultural Heritage Foundation, and the Federal Commissioner for the Records of the State Security Service of the former GDR are italicized. These quotations are either available to the author as files or are taken from the book "Versagtes Vertrauen. Wissenschaftler der DDR im Visier der Staatssicherheit" by R. Buthmann, as indicated. Quoted words, unless they refer to proper names, are reproduced verbatim from the cited publications, preserving the original spelling. Group photos distributed free of charge to all participants of scientific conferences are considered to be in the public domain. Unless otherwise stated, photographs were taken by the author or are in the public domain (gemeinfrei). References in footnotes are kept as concise as possible and serve merely as orientation codes for the final list of references.

I would like to thank the Potsdam photographers Blumrich, Fahlbusch, Rohn, and Schulz-Fieguth, who have generously shared their valuable collections with me over the years. I am also especially grateful to Kurt Arlt and Jürgen Rendtel from the Förderverein Großer Refraktor Potsdam e.V., and to Peter Kroll from the Sternwarte Sonneberg, for providing photos from their archives. My thanks also go to Isolde Meinunger from Steinach for granting access to her personal collection. The geophysicist Franz Jacobs kindly provided me with documents from his work on the evaluation commission of the Institutes for Geo- and Cosmic Sciences of the German Science and

© The Author(s), under exclusive license to Springer Nature Switzerland AG 2026
G. Rüdiger, *The Astrophysical Observatory Potsdam - Triumph and Tragedies*, Astronomers' Universe, https://doi.org/10.1007/978-3-032-06294-9_11

Humanities Council. Melissa Thies from the library of the Leibniz Institute for Astrophysics Potsdam (AIP) deserves special thanks for her constant and prompt support. A big thank-you also goes to Anna Krutsch and the Deutsches Museum in Munich for their always quick and invaluable assistance. Noam Libeskind from the AIP deserves special thanks for reviewing the final text for linguistic accuracy. Last but not least, without the ongoing support of my historically experienced colleagues Wolfgang R. Dick (Potsdam) and Reinhard E. Schielicke (Jena), I might have had to abandon the project halfway.

Appendix

Appendix 1 NS Official Section of the NSDAP, Fachschaft of Observatories, Potsdam District Group[1]

Since 1 June 1920, Professor Erwin Freundlich has been working at the Astrophysical Observatory as a Prussian official, initially as an observer and now as a Principal observer. Prof. Freundlich's father was Jewish. Before war, and again now, he signed himself unjustifiably as "Finlay-Freundlich" by prefixing his name with that of his maternal grandfather, a famous English astronomer and comet discoverer. During the war, the English word "Finlay" was missing from his name, which only reappeared after the war.

At the beginning of 1932, Professor Freundlich was appointed director of the Einstein Institute, now the Institute for Solar Physics, which was affiliated to the Astrophysical Observatory, by the November party's Minister for Science, Art and National Education. At the same time, in a deliberate humiliation of the nationalist director, Professor Ludendorff it was given him an unusual degree of independence, both in the management of the scientific work of the Einstein Institute and in administrative matters, which is not usual for section heads.

---

[1] Subformation of NSDAP. Translated from Gruner (2008), document 63, p. 210; letter dated 18 July 1933, signed by Fachschaftsleiter Ernst Obst, addressed to the NS-Beamtenabteilung des Gaus Kurmark der NSDAP, Berlin.

G. Rüdiger, *The Astrophysical Observatory Potsdam - Triumph and Tragedies*, Astronomers' Universe, https://doi.org/10.1007/978-3-032-06294-9

Instead of protecting the nationally-minded director against the impertinence and interventions of his Principal observer, the November government placed the anti-nationalist Jewish descendant on an almost equal footing with the director.

However, this equalisation has since been largely reversed by the National Socialist government. This did not, however, eliminate Prof. Freundlich's un-German character. As proof of his un-German behaviour, a statement by the Principal Observer Prof. Dr. Münch is enclosed. In addition, I have in my hand written information from recent times which I received through official channels. They show that Freundlich has made statements about leaders of the NS movement which must deeply outrage every German citizen. From all of this, the NS Fachschaft of the observatories has come to the conclusion that Prof. Freundlich is not suited to be an official in the new Reich, and least of all a civil official in a leading position, due to his anti-people attitude. It is doubtful whether a dismissal in accordance with §3 of the Law for the Restoration of the Professional Civil Service is applicable to Prof. Freundlich. On the other hand, the Fachschaft considers dismissal from the service or at least removal from a leading position to be necessary, as he offers absolutely no guarantee that he will always stand up for the national state without reservation. On the contrary, the Fachschaft has gained the impression that he hates the national movement.

The observatories' Fachschaft therefore requests the Gauleitung Kurmark to pass on the above information to the relevant authority.

Appendix 2 Nobel Laurates support Erwin Freundlich[2]

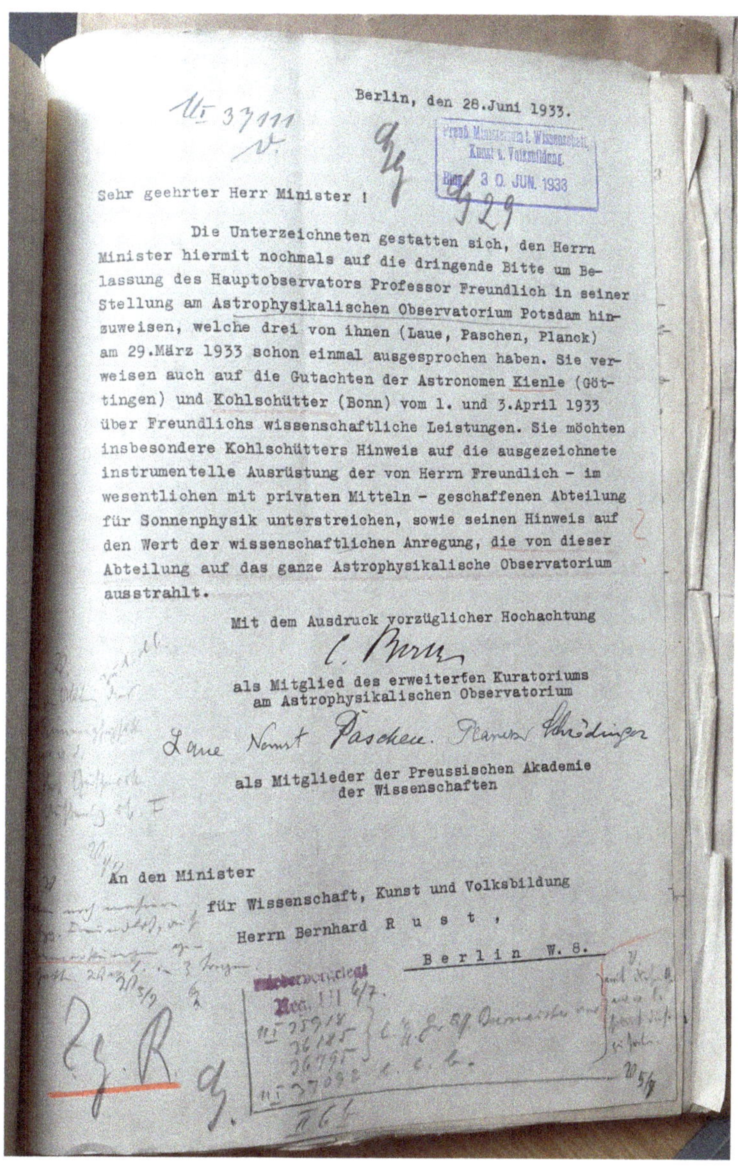

---

[2] Letter from Bosch, Laue, Nernst, Paschen, Planck and Schrödinger dated 28 June 1933 to the Minister of Science, Art and National Education Bernhard Rust. GStAPK, I HA, Rep. 76, Vc Sekt. 1, Tit. XI, Part II No. 6b, Vol. 10. See p. 65.

Appendix 3 Autobiography of Harald v. Klüber[3]

Born the 6th of September 1901 at Potsdam (Germany) as only child of the than Oberleutnant at the Jäger zu Pferde Robert von Klüber (1873–1919), protestant, and of his wife Elsa, née von Mühlberg (1877–1945).

[…]

Grown up at Potsdam, Berlin, for a short period in Bruxelles (Belgium) (where his father was at the staff of the German embassy) and again in Berlin. For many years most happy holidays at the castle of Georgenthal near Gotha (Thüringen) at the home of the grandparents von Mühlberg.

During the first years private education because of the frequent change of domicile of the parents (father officer and in diplomatic service), later at school at the Falk-Realgymnasium in Berlin. Maturity (Abiturium) in 1920. From the age of ten on clear and strong interests in Technics, Physics and Astronomy, with many own activities in these fields. The father, in high leading military positions during the war 1914–1918 and decorated with the order Pour le Mérite, lost his life 1919 during operations for the protection of the German National Assembly at Weimar.

From Easter Terms 1920 to 1924 at the University and at the Technische Hochschule in Berlin, attending lectures on Physics (M. Planck, H. Rubens, v. Laue, W. Nernst, A. Einstein, P. Pringsheim), Mathematics (L. Bieberbach, E. Schmidt, v. Mieses), Astronomy, Astrophysics (P. Guthnick, F. Cohn, A. Kopff, G. Witt) and Philosophy (cognition, history of philosophy, indian philosophy, Kant, Schopenhauer). Interests and talents predominantly in Basic Research, especially in the line of laboratory experiments and in observations. During holidays and in spare time very studious in order to achieve a wide and general knowledge in many other subjects. 1924 degree of Dr. phil. at Berlin University by a paper dealing with astronomical observations carried out at the observatory of the Technische Hochschule (A. Miethe) at Berlin and winning the certificate Valde laudabile.

Since 1923 during the last university term attached to the recently founded Einstein-Institut in Potsdam (Prof. E. F. Freundlich) in the grounds of the Astrophysical Observatory at Potsdam. Working on a research program concerning high ionized emission lines in an electric arc of highest energy for a possible explanation of solar corona emission lines. These investigations for

---

[3] Excerpt of the Potsdam-related part of the full text given in Rüdiger (2024). Source Landesarchiv Baden-Württemberg, Generallandesarchiv Karlsruhe, Findbuch 69, Nachlass von Klüber, No. 37. The title, Klüber's minor translation errors and the sometimes incorrect spelling have been retained, The transcription was made by the author and proofread by W. R. Dick.

the Einstein-Institut were carried out at the large research laboratories of the Siemens-Concern at Berlin-Siemensstadt.

Since 1924 permanently attached to the Einstein-Institut in Potsdam. 1924 scientific assistant, 1933 Observer, 1946 Chief-Observer and Head of the solar department, appointed Professor 1941. For principal reasons and in spite of being a scientific civil servant never member of the Nazi-Party or of any of its active organisations. 1935 purchase of a small house at Potsdam, Finkenweg 6.[4] Hobbies: water- and motorcarsport, photography, archaeology, history of art of ancient cultures, tape recording, biology.

1924–1926 extensive work for final adjusting and bringing into operation the solar tower telescope (Einsteinturm) at the Potsdam observatories, an instrumental type so far nearly unknown in Europe. During the same period preparation for and afterwards participation in the Potsdam solar eclipse expedition to Benkoelen (South Sumatra) for testing the light deflection according to Einstein's general theory of relativity during the total solar eclipse of 1926, January 14th (leader Prof. E. F. Freundlich). Returning from Sumatra via the United States and after having obtained a special grant as a scientific visitor v. Klüber stayed at the Bosscha-Sterrenwacht at Lembang (Java), the Tokyo Observatory, the Lick- and the Mt. Wilson Observatories.

1927–1930 once more thorough preparations for and then participation in the large second Potsdam eclipse expedition to Takengon (North Sumatra) for testing once more Einstein's Light Deflection during the total solar eclipse of May 9th, 1929. Special responsible for a large part of the complicated adjustment work, for all photographic work, for the time service and for operating one of the major instruments. Since observations were very successful v. Klüber was stationed with the instruments in Takengon till January 1930 in order to obtain the necessary comparison observations. During this period extensive travels were carried out through all of Sumatra, through Java (Bosscha Sterrenwacht) and Bali. This expedition and the prolonged stay in Takengon and Indonesia are an especially pleasant recollection. 1930 return journey with longer interruptions in Malaya, Burma, Siam, Indochina, North-India and Aegypt. 1930–1932 mainly occupied with the time consuming measuring and reduction work of the very good plates obtained during that expedition.

At about 1927 first tentative pioneer work concerning microphotometric investigations of Fraunhofer lines were carried out using the very first prototype of a newly constructed selfrecording photoelectric photometer of high precision. This work was later continued.

---

[4] Constructed 1926/1927.

During this time also erection, near the Einstein tower, of the 50-cm reflecting Goerz telescope transferred from the Technische Hochschule at Berlin; later on, photographic work in stellar statistics was carried out with this instrument as well as with the 20-cm Zeiss astrograph, also in operation near the solar tower.

During this and the following years a number of papers was published on astronomical instruments and on history of astronomy.

Since 1927 permanent scientific advisor of the Askania Factory (precision optics and instruments). 1924–1945 fellow of the Astronomische Gesellschaft and since 1933 fellow of the Royal Astronomische Gesellschaft, London. Since 1934 astronomical collaborator of the Frankfurter Allgemeine Zeitung, where many articles were published.

The small group of scientists originally assembled and working at the Einsteinturm since 1924 (Freundlich, von der Pahlen, Grotrian, v. Klüber, Feinmechaniker Strohbusch) proved for many years to be a most happy team. The writer is especially obliged for many reasons to Professor Freundlich during these years of their mutual collaboration. Remembering this time, the writer is convinced that in this period of fast advancing astrophysics he hardly could have found a more attractive and instructive position in Germany than in this lively scientific and extremely interesting and pleasant atmosphere around the Einsteinturm. Contacts could be made there with many of the leading scientists of these days (i.g. Carl Bosch, Eddington, Einstein, v. Laue, Lyot, Milne, Minnaert, Nernst, Planck, Sommerfeld and many others).

In spite of being bound to serve in the army v. Klüber actually never was called to any military service. During the war (1939–1945) he first was employed like many other German colleagues in extensive calculations for astronomical navigation. Later on, he was nearly exclusively occupied with basic research in practical solar physics and its relation to the formation and structure of the ionosphere and to short wave radio transmission.

Since 1941 the writer has started a lengthy project, meant to be extended over many years, on the systematic investigation of solar magnetic fields with the powerful optical equipment of the Potsdam towertelescope. This field of investigation was till then nearly an exclusive privilege of the Mt. Wilson Observatory. Since 1949 this kind of observation was carried on by v. Klüber after having moved to Cambridge and later to Malta (1965–1971). When still in Potsdam (1943) interferometric methods with very high resolving power, till then hardly used in solar spectroscopy, were developed and applied successfully by v. Klüber to the investigation of Fraunhofer lines and solar Zeeman effects.

The solar installation at Potsdam suffered very heavy damage during an air raid on 1945, April 14th. Nevertheless, repair work was started immediately and with the effective support of the sowjetic occupation force scientific research could soon be resumed again.

In 1948 v. Klüber found himself attached for a year to the Observatory of the Eidgenössische Technische Hochschule at Zürich, Switzerland. There he had the long desired opportunity to work for most of the year at the High-Altitude Station at Arosa. Predominently systematic observations with the Lyot coronagraph were carried out there. Also the new large horizontal solar telescope, build by Grubb-Parsons, Ltd., England, was installed and adjusted at Arosa during that period.

....

Appendix 4 H. Alfvén to W. Grotrian

KUNGL. TEKNISKA HÖGSKOLAN   Stockholm den 11. Oktober 1950

INSTITUTIONEN FÖR ELEKTRONIK

STOCKHOLM 26

TELEFON: VÄXEL 23 28 20

HA/Hn

Herrn

Professor W. Grotrian

Deutsche Akademie der Wissenschaften zu Berlin

Astrophysikalisches Observatorium

Potsdam

Sehr verehrter Herr Kollege!

Mit vielem Interesse habe ich Ihren Brief studiert. Es ist ganz richtig, dass ich auch selbst nicht sehr zufrieden bin mit dem Erregungsmechanismus der Sonnenflecken, und meine Darstellung muss als einen ersten Versuch betrachtet werden. Die besten Stützen für meine Auffassung sind:

1. Aus dem Sonnenfleckenmaterial kann man "sunspot progression curves" herleiten (p. 120), die viel regelmässiger sind als die meisten anderen Fleckenerscheinungen. Ich glaube, dass dies so gedeutet werden muss, dass man hier eine Grösse gefunden hat, die nur von der Geometrie des allgemeinen Sonnenfeldes herrührt.

2. Auch glaube ich, dass die Korrelation zwischen den beiden Hemisphären (p. 188) eine gute Bestätigung der Grundlagen darstellt. Ich habe gerade dieselbe Untersuchung zur 18. Sonnenperiode erweitert und nach vorläufigen Ergebnissen wiederholt sich dieselbe Kurve auch für diese Periode.

Mit diesem Observationsmaterial ist es mir sehr schwer zu glauben, dass das allgemeine Sonnenmagnetfeld variabel sein kann. Zwar muss natürlich das Feld immer durch Wellenbewegungen von verschiedener Grösse überlagert sein, aber der Grundriss muss doch unverändert bleiben, sonst könnte man nicht die grosse Regelmässigkeit der "sunspot progression curves" erklären. Es ist nicht ganz notwendig, dass das Feld auch in der Photosphäre ganz regelmässig ist, sondern man könnte sich denken - obwohl mit beträchtliche Schwierigkeiten - dass die Zeemaneffektmessungen doch nichts wesentliches zu unser Kenntnis vom Sonnenmagnetfeld geben. Die grosse experimentellen Schwierigkeiten und die sehr widersprechenden Ergebnisse des Messungen geben mir wenig Vertrauen zu diesen Messungen überhaupt. Ich glaube das heute das beste Wert des allgemeinen Magnetfeldes durch die "sunspot progression curve" gegeben ist (p. 124).

Mit kollegialen Grüssen

Ihr sehr ergebener

H. Alfvén

[...] I have studied your letter with great interest. It is quite correct that I am not entirely satisfied with the excitation mechanism of sunspots myself, and my presentation must be regarded as a first attempt. The best supports for my view are:

1. From the sunspot material, one can derive "sunspot progression curves" (p. 120), which are much more regular than most other sunspot phenomena. I believe this must be interpreted as having found a quantity here that is solely determined by the geometry of the general solar field.
2. I also believe that the correlation between the two hemispheres (p. 138) provides good confirmation of the explanation. I have just extended the same investigation to the 18th solar period, and according to preliminary results, the same curve is repeated for this period as well.

With this observational material, it is very difficult for me to believe that the general solar magnetic field can be variable. Of course, the field must always be superimposed by waves of various sizes, but the basic structure must remain unchanged; otherwise, one could not explain the great regularity of the "sunspot progression curves". It is not entirely necessary for the field to be completely regular in the photosphere either, but one could imagine—although with considerable difficulties—that the Zeeman effect measurements do not provide anything essential to our knowledge of the solar magnetic field. The great experimental difficulties and the very contradictory results of the measurements give me little confidence in these measurements at all. I believe that today the best value of the general magnetic field is given by the "sunspot progression curve" (p. 124).

With collegial regards,
Yours sincerely,
H. Alfvén

Appendix 5 Succession Grotrian 1954–1956[5]

-- W. Friedrich to H. A. Brück, 31 May 1954

[…]

Professor Grotrian's death has deprived the Astrophysical Observatory of its director. The German Academy of Sciences in Berlin is faced with the task of finding a successor who will continue the high scientific tradition of the Potsdam Observatory and ensure that the research facilities that have been created there since 1946 and will be expanded in the future are fully utilised. In recent years, considerable investments have been made to modernise the instrumental equipment of the Astrophysical Observatory, and the radio astronomy section has received its first working instrument. A 2mmirror telescope is planned for the Astrophysical Observatory. In view of the unusual scientific importance of the Astrophysical Observatory, the Academy has also considered scholars from abroad when looking for suitable personalities of international renown.

In an effort to avoid wasting valuable time in negotiations with candidates who are not willing to seriously consider accepting a call, the Academy asks for your understanding for the perhaps somewhat unusual form of a simultaneous enquiry to all gentlemen under consideration.

The German Academy of Sciences and Humanities in Berlin asks you […] to answer the following questions:

1. Would you yourself be prepared in principle to accept an appointment to the German Academy of Sciences as Director of the Astrophysical Observatory Potsdam?
2. Which astronomers do you consider suitable to take on such a task?

[…]

-- H. A. Brück to W. Friedrich, 25 August 1954:

[…]

Thank you very much for your kind letter of 29 July, which I received on the eve of a trip to Germany. On this trip I spoke, among others, to Prof. Kienle in Heidelberg and in this way learnt a number of things about the current research possibilities at the Astrophysical Observatory in Potsdam. Unfortunately, this time it was not possible for me to come to Berlin, where many gentlemen were on holiday.

---

[5] ABBAW, Abtl. Akademiebestände nach 1945, AKL (1945–1968), No. 13. Translations from the German originals.

Dunsink Observatory,
Co. Dublin.

25. August 1954

Herrn Prof. Dr. W. Friedrich,
Präsident der Deutschen Akademie der Wissenschaften
zu Berlin .

Sehr geehrter Herr Präsident,

                    Sehr vielen Dank für Ihren
liebenswürdigen Brief vom 29. Juli, den ich am Vorabend einer Reise
nach Deutschland erhielt. Auf dieser Reise habe ich unter anderem
Herrn Prof. Kienle in Heidelberg gesprochen und auf diese Weise
eine Reihe von Dingen über die augenblicklichen Forschungsmöglichkeiten
auf dem Astrophysikalischen Observatorium in Potsdam erfahren. Leider
war es mir dies Mal zeitlich nicht möglich, nach Berlin zu kommen, wo
wohl überdies viele Herren in den Ferien waren.

    Wie ich Ihnen bereits schrieb, würde ich gern die Situation an
Ort und Stelle sehen und dabei einige für meine Entscheidung wichtige
Fragen durchsprechen. Ich danke Ihnen deshalb sehr für Ihre liebens-
würdige Einladung. Wie ich bereits Herrn Prof. Wempe gesagt habe, und
soweit ich die Dinge im Augenblick übersehen kann, würde mir die Zeit
um den Anfang Oktober am besten passen. Ich habe hier im Laufe des
September einige recht wichtige Sitzungen, deren genaue Daten noch
nicht festgelegt sind. Ich glaube aber, dass ich Ihnen bis spätestens
Mitte September mitteilen kann, wann ich nach Berlin kommen könnte.
Ich würde Ihnen wenigstens drei Wochen vor meinem Besuch schreiben
und nehme an, dass dies genügend Zeit geben würde, ein Visum zu
erhalten. Falls dies nicht rechtzeitig genug ist, haben Sie vielleicht
die Freundlichkeit, es mich wissen zu lassen.

        Mit dem Ausdrucke meiner vorzüglichsten Hochachtung
                verbleibe ich
                Ihr sehr ergebener

                H. A. Brück.

As I have already written to you, I would like to see the situation on the
place and discuss some important questions for my decision. I would therefore like to thank you very much for your kind invitation. As I have already
told Prof Wempe [...], the time around the beginning of October would suit
me best. [...] I would write to you at least 3 weeks before my visit and assume
that this would give me enough time to obtain a visa. [...]
-- M. Minnaert to W. Friedrich, 22 June 1954:
[...]

Personally, I would very much honour an appointment as director of the observatory in Potsdam, but would not accept it because I consider myself too connected to my current work and my environment.

The question of which astronomers would be suitable is very difficult to answer. It seems to me that it is necessary to find a man of great scientific merit who also has a certain understanding of the recent development of your country. There is no point in naming names of capable astronomers who, for general reasons, would not accept a call to East Germany.

[Three suggestions for experienced astronomers and the names of three inexperienced ones follow].

I sincerely hope that the Academy will succeed in making a good choice and thereby promote the development of the Astrophysical Observatory. I would like to express my sincere gratitude for the honour you have done me by sending me your letter.

[...]

„Ich hoffe aufrichtig, dass es der Akademie gelingen wird eine gute Wahl zu treffen, und dadurch die Entwicklung des Astrophysikalischen Observatoriums zu fördern. Für die Ehre welche Sie mir durch Sendung Ihres Briefes getan haben, bezeuge ich Ihnen meinen verbindlichsten Dank.
Mit vorzüglicher Hochachtung,
M. Minnaert

-- M. Schwarzschild to W. Friedrich, 12 July 1954:

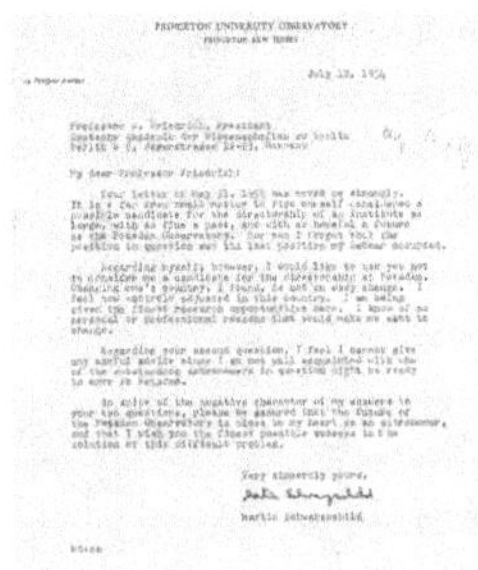

-- H. v. Klüber to H. Kienle, 30 January 1956:

[...]

Thank you for your kind letter of 25 March, which, as you can see, I first had to think about a little.

You may know that our former Potsdam Institute is closer to my heart than anyone else's; not only have I spent most of my academic life there, but Potsdam is also my home town. It is therefore only natural that I would love nothing more than to return there. For the same reason, I also feel a very strong obligation to be available to the Observatory and its important tradition when it is necessary, as it is now, and no one better can be found. The excellent conditions offered in Potsdam are just as important as personal wishes and even my own security. Incidentally, I view the position in Potsdam from a much higher level than that of our particular academic discipline; I see it as an important cultural-political position in the unfortunate conflict between East and West, which must somehow be resolved in a reasonable manner.

For similar reasons, I strongly advised Mr. Brück, who recently spent several hours with me, to accept Potsdam, because I consider him to be a particularly suitable man under the given circumstances. I very much regret that he declined for obviously personal reasons, which I do not quite approve of.

But my own opinion is something like this:

You yourself know best that at the end of the war and although my personal and family interests were already completely in the West at that time, I remained in Potsdam with the firm intention of fulfilling my duty at the institute even in bad times. I was also determined to co-operate loyally with the new rulers in the long term and to the limit of what was possible. However, I was not then and am not now willing to abandon certain moral principles to which I feel fundamentally bound. That's why I believe that working in the GDR with its particular ideology, especially in a manager position, would very quickly entangle me in conflicts that could ultimately only be detrimental to the cause and the Institute.

As you know, I had the opportunity to work with the Potsdam expedition group for several months this year as leader of the large English expedition to Ceylon. I was particularly pleased with the group's solid scientific preparation and work, which still met the old standards of our Potsdam Institute. And I also gained the particularly impressive impression that it is simply a pity how a whole group of ambitious young people in Potsdam today are apparently muddling around in the dark without proper leadership. I can't even imagine who could be brought in as a leader. My hope, for years now, has always been Minnaert but he's probably no longer available. Have they thought about Houtgast been considered? A foreigner would have a relatively strong position in the GDR.

Incidentally, it would be very welcome if you would take the opportunity to let the Academy of Sciences in Berlin know that it can only be noted with the greatest disapproval that they not only employ former Nazis from the intelligentsia (who must ultimately be regarded as fully responsible for their behaviour) in office and dignity, but that such people, for example in our field, are increasingly being sent abroad as representatives of the GDR. This accusation is already serious enough in the case of the Federal Republic, but in the case of the GDR it seems absolutely inexcusable.

You are right that we are happy here [...]. [Details of the academic conditions and life in Cambridge follow].

Appendix 6 P. H. Roberts to H. Stiller

THE UNIVERSITY OF NEWCASTLE UPON TYNE
[...]
18th April, 1969.
Dear Professor Stiller,
My colleagues and I from this department greatly enjoyed the participation of your colleague, Dr. Krause, in the conference held here from April 14th to April 18th and jointly organised by the School of Mathematics and Physics in this University. We valued his contributions to the proceedings. I would like to explore with you the possibility of further strengthening links between your Institute and our department through our common interests in dynamo theory. I have in mind the thought that Dr. Krause (or one of his associates) might spend a few weeks here next academic year (i.e., between October 1969 and June 1970) and that, reciprocally, I (or possibly one of my group) could spend a comparable time at your Institute. Of course, until I hear your reaction to this proposal, I will not attempt to go into details nor will I try to set the necessary University machinery into motion, but I would imagine that it would be left to Dr. Krause and myself to obtain travel funds from our respective organisations, and that each organisation would provide adequate funds to cover the living expenses of the visitor from the other. We would also each attempt to ease the others visa formalities.

Yours faithfully,
P. H. Roberts
Professor of Applied Mathematics
Professor Stiller,
Director,
Central Institute Physics of the Earth,
Potsdam 15, Telegrafenberg,
Germany, D.A.R.

Appendix 7 IM Martin: Situation at ZIAP[6]

XIII/Inst. Potsdam, d. 05.07.1975

The difficulties at the ZIAP begin with the administration of the institute. Prof. Treder, who is inexperienced in management issues and has never attended a management meeting, sporadically tries to take control of the administration. However, this desperate attempt must fail. Naturally, this creates considerable difficulties for the managing director. On the other hand, no one attempts to reassign Prof. T., to give him a working group and let him work. The fear of T.'s connections at the Central Committee of the SED is too strong.

In the past, the IM held several discussions with **XXXX** regarding the political and ideological situation in the ZIAP. The conclusion of the discussions is

* Prof. T. will no longer interfere in the management of the ZIAP
* Dr. Rüdiger and Dr. Rädler will not habilitate
* Prof. T. is in agreement with the dismantling of working group Dr. Krause.

The IM is of the opinion that with a combination **XXXX**/Ruben (theory/observation), the ZIAP could be healthily downsized. The question of location would also have to be discussed again, as the observation centres had all moved to the south of the GDR. Furthermore, the IM did not rule out the possibility that a high state medal had been requested for Prof. T. by the President of the AdW of the GDR.

Signatures

Appendix 8 Instruction on the Selection, Confirmation and Preparation of Travelling and Foreign Cadres[7]

* §3.1 The directors and heads of the institutions are responsible for ensuring that only those persons are selected as travelling and foreign cadres whose political reliability has been proven and who represent the GDR abroad in a worthy manner.

---

[6] BStU, translated copy from the archive of the author. IM "Martin" = H. Stiller, Head of Department Geo- and Cosmos Sciences of the AdW since 1973.

[7] Extracts. Translated instruction 21/82 of 18 June 1982 (classified information), here: Draft of 16 March 1982 Source BBAW AKL (1969–1991), No. 853.

- §4.1 The directors and managers of the organisations are obliged to discuss every cadre proposal in the advisory and control group for the comprehensive assessment of the cadres.
- §4.2 Cadres shall not be informed of their deployment until they have received the approval of the competent office of the Ministry of State Security within the scope of its security policy responsibility.
- §4.3 The counselling and decision on the confirmation of a cadre is made on the basis of a [decision] document to be drawn up. It consists of a personal data sheet, a list of relatives of the spouse, children, parents, siblings, parents and siblings of the spouse and an assessment of the cadre.
- §4.4 With a written application for approval as a travelling or foreign cadre, the decision document must be handed over to the responsible employee of the Ministry of State Security. The director or head of the organisation shall be informed of the security policy decision of the Ministry of State Security when the decision document is returned. Only then can the application be handed over to the Director of cadre and education for confirmation and registration.
- §4.5 It is not permitted to invoke the security policy decision of the Ministry of State Security to justify the rejection of a cadre or the withdrawal of confirmation by the Director of cadre and education, his authorised representatives or members of the advisory group with regard to travelling and foreign cadres.
- §4.9 Family members are only permitted to travel with cadres on assignments abroad lasting longer than 12 months.
- §5.2 The advisory group is responsible for providing comprehensive support to the director or head of the institution in the selection, confirmation and preparation of travelling and foreign cadres and for making recommendations for his or her own decisions.

Appendix 9 Military Mission[8]

[...]

1. during your stay in Helsinki, prioritise people who are seeking contact with you and record their contact interests.
In principle, you should be open and interested in contact attempts and show your interest in the person in question by asking skilful and targeted questions in personal conversations.
2. in the event of provocations regarding co-operation with the MfS that cannot be ruled out, always assume that your conspiracy is guaranteed by the MfS. On this basis, vigorously reject such provocations and only name your official contacts with the Institute's security officer.
3. specific information requirements
During your stay abroad, work out tips according to your possibilities

- on identified military objects, military personnel, military technology and military movements,
- on questions of the border regime,
- for overnight stays in the accommodation allocated to you,
- on the mood of the population,
- on political events and evaluations.

During your stay abroad, no recordings will be made of you that could provide information about the assignment given to you by the MfS.

[...]

---

[8] Translated order from the MfS dated 12 September 1985 for a business trip to the University of Helsinki. BStU-copy in the archive of the author.

<u>A u f t r a g</u>

Im Zeitraum vom 24.08.85 bis 04.10.85 werden
Sie zu einem Arbeitsaufenthalt an der Universität Helsinki(Dr. Ilkka Tuominen) weilen.
Im Zusammenhang mit dieser Dienstreise erteilt ihnen das
MfS folgenden Auftrag:
1. Während Ihres Aufenthaltes in Helsinki beachten Sie vorrangig Personen, die zu Ihnen Kontakt suchen und erfassen deren Kontaktinteressen.
Gegenüber Kontaktversuchen zeigen Sie sich prinzipiell
aufgeschlossen und interessiert und lassen in persönlichen
Gesprächen auch Ihr Interesse an der jeweiligen Person
durch geschickte und gezielte Fragen erkennen.

2. Bei nicht auszuschließenden Provokationen hinsichtlich
einer Zusammenarbeit mit dem MfS gehen Sie grundsätzlich
davon aus, daß Ihre Konspiration seitens des MfS garantiert
ist. Weisen Sie auf dieser Basis entsprechende Provokationen energisch zurück und benennen Sie lediglich Ihre offiziellen Kontakte zum Sicherheitsbeauftragten des Institutes.

3. Spezifischer Informationsbedarf
Erarbeiten Sie während Ihres Aufenthaltes im Ausland entsprechend Ihren Möglichkeiten Hinweise

- zu festgestellten militärischen Objekten, Militärpersonen,
  Militärtechnik und militärischen Bewegungen,
- zu Fragen des Grenzregimes,
- zur Übernachtung in Ihnen zugewiesenen Unterkünften,
- zur Stimmungslage der Bevölkerung,
- zu politischen Ereignissen und Bewertungen.
Während des Aufenthaltes im Ausland werden von Ihnen keine
Aufzeichnungen angefertigt, die Aufschlüsse über den Ihnen
vom MfS erteilten Auftrag geben können.
Nach Rückkehr von Ihrer Dienstreise melden Sie sich vereinbarungsgemäß am.............. um............. Uhr am festgelegten
Treffort.

Potsdam, den 12.09.85   Schmall

Appendix 10 Fritz Krause (1927–2024)[9]

Gustav Emil Fritz Krause passed away on February 28, 2024 in Wilhelmshorst near Potsdam. More than 60 years ago, he presented his first talk "A dynamo model for the 22-year cycle of the solar magnetic field" at IAU Symposium No. 22 in Munich to the astronomical community, which had initially been extremely sceptical about the turbulence theory approach developed in Jena for the formation of cosmic magnetic fields.

Fritz Krause came from the early mathematics school in Jena and attended physics lectures by Friedrich Hund. After completing his doctorate, he joined Max Steenbeck at the nearby Academy Institute for Magnetohydrodynamics in 1958, where he developed the mathematical foundations of the dynamo theory of cosmic magnetic fields. The publication "Berechnung der mittleren Lorentz-Feldstärke $\langle \boldsymbol{u'} \times \boldsymbol{B'} \rangle$ für ein elektrisch leitendes Medium in turbulenter, durch Corioliskräfte beeinflußter Bewegung" with M. Steenbeck and K.-H. Rädler in Zeitschrift für Naturforschung in 1966 was by far the internationally most successful journal publication in the history of science in the GDR.

Krause arrived at the Friedrich Schiller University in Jena at the age of 18 as a result of the events of the WWII. Born the youngest in Groß Särchen in

---

[9] G. Rüdiger, Astron. Nachr. 345, e240048 (2024).

Niederlausitz, which is now part of Poland, the Krauses had to leave their homeland in 1945 and reached Friedrichroda in Thüringen as a refugee family. Fritz studied mathematics in Jena from as early as 1946, although he was allowed to take his university entrance qualification retrospectively in the second year at a special FSU institution. Becoming a biologist was another career option which remained present in many of his later hobby activities. He completed the mathematics degree in 1951 under Walter Brödel,[10] in whose Institut für Reine Mathematik he also wrote his dissertation "Zur konformen Geometrie der dreifachen Orthogonalsysteme". In 1958, the year he graduated, Irmgard Schröder and Fritz Krause started a family, which soon included their sons Matthias und Peter.

After a brief period as director of the Geomagnetisches Institut der Akademie der Wissenschaften zu Berlin on Potsdam/Telegraphenberg, Krause became head of the "Kosmische Magnetfelder" section at the Astrophysikalisches Observatorium Potsdam, later at the Zentralinstitut für Astrophysik, where hydrodynamic and thermodynamic applications were also developed. The original German-language publications from Jena were translated into English by P. H. Roberts and M. Stix soon after they appeared and have since become one of the cornerstones of the new scientific branch of dynamo theory. Some of the terms introduced at that time, such as α effect or helicity, are still used today in modern publications. Together with K.-H. Rädler, he wrote the widely acclaimed monograph "Mean-field magnetohydrodynamics and dynamo theory" in 1980, the Observatory's first book publication since the end of WWII. In the 70s, we were often visited by Paul H. Roberts from University of Newcastle upon Tyne—who could hardly be beaten at table tennis—also in order to break through the isolation of the Potsdam staff. That also was the time when both were involved in ongoing scientific disputes as a team. The first international conference on the Telegraphenberg for decades, "Stellar and planetary magnetic fields" in 1983, was a final stage in the observatory's rebirth—although the AOP disappeared shortly afterwards as a result of untransparent decisions from the outside.

At the new workplace in Sternwarte Babelsberg, the stellar physics aspect of the dynamo group was lost, partly because the Max-Planck-Institute for Radio

---

[10] Brödel was West German citizen and taught in Jena mathematics starting 1947 under permanent opposing by communistic activists. After his refusal of the Berlin Wall, the SED district leadership launched a public campaign: "How much longer will a NATO professor be teaching at our Friedrich Schiller University?" In December 1961 he was banned from the university, left the GDR and was subsequently dismissed under withdrawal of the title of professor. Steenbeck also took part in this campaign with all his authority and demanded a "clear line between Mr. Brödel and the University." In the spring of 1990, Brödel was happy to re-establish the contacts to his old friends at the institute of mathematics in Jena.

Astronomy in Bonn was currently carrying out sensational measurements of magnetic fields on a large scale. "We now can see the toroidal fields we have always talked about", exclaimed Krause at the 1988 workshop on "Magnetic fields in galaxies" in Babelsberg, which was organized jointly with Wielebinski's group in Bonn. The galactic magnetic fields and their symmetries—axisymmetric or nonaxisymmetric—became the preferred testing objects of the global dynamo community at this time. Their observations and theories also dominated the IAU Symposium No. 157 "The cosmic dynamo", which Krause organized virtually by himself and which was terminated to coincide with the 100th anniversary of the death of the inventor of the dynamo principle, Werner von Siemens. It was the first major event worldwide dedicated solely to the self-excitation of large-scale cosmic magnetic fields and was the first to unite Krause's various circles of scientific friends without any political restrictions.

In one of his last publications in retirement, which he reached with this conference in 1992, Krause attacked one of the most neglected problems of dynamo theory. The magnetic quadrupole solutions observed in galaxies require tiny seed fields of the same symmetry for self-excitation, but their generation by contraction in the vertical direction is difficult to imagine.

On his 80th birthday, the Leibniz-Institut für Astrophysik Potsdam honored Fritz Krause—also a member of the Deutsche Akademie der Naturforscher Leopoldina since 1980—with a special colloquium with main contributions by R. Beck (Bonn), J. Kurths (Berlin) and F. Stefani (Dresden-Rossendorf). He spent the last years of his life in the Seniorenresidenz St. Elisabeth in Wilhelmshorst, where he died surrounded by his family shortly before reaching the age of 97.

References

Alfvén, H. (1941). On the solar corona. *Arkiv för Matematik, Astronomi och Fysik, 27*, 1.

Alfvén, H. (1950). *Cosmical electrodynamics*. Clarendon Press.

Arlt, K. (2007). 175 Jahre Telegraphenberg. *Mitteilungen Studiengemeinschaft Sanssouci e.V., 12*, 8.

Arlt, R., & Vaquero, J. M. (2020). Historical sunspot records. *Living Reviews in Solar Physics, 17*, 1.

Auth, J. (1982). Albert A. Michelson an der Berliner Universität. *Astronomische Nachrichten, 303*, 7.

Baker, N. (1965). Periods and pulsational stablity of RR Lyrae stars. *Kleine Veröffentlichungen der Remeis-Srernwarte Bamberg Bd. IV, Nr. 40*, 121.

Beck, R. (1982). The magnetic field in M31. *Astronomy and Astrophysics, 106*, 121.

Beck, R., et al. (1987). The magnetic field in M51. *Astronomy and Astrophysics, 186*, 95.

Becker, F. (1929). Spektral-Durchmusterung der Kapteyn-Eichfelder des Südhimmels. *Publikationen des Astrophysikalischen Observatoriums zu Potsdam, 27*, 5.

Becker, W. (1942). *Sterne und Sternsysteme*. Verlag Theodor Steinkopff.

Behr, A., & Siedentopf, H. (1952). Zur Statistik von Sonneneruptionen. *Zeitschrift für Astrophysik, 30*, 17.

Belopolsky, A. (1893). Über die Sonnenrotation aus Fackelpositionen. *Astronomische Nachrichten, 132*, 213.

Berdyugina, S., & Tuominen, I. (1998). Permanent active longitudes and activity cycles on RS CVn stars. *Astronomy and Astrophysics, 336*, 117.

Biermann, L. (1951a). Bemerkungen über das Rotationsgesetz in irdischen und stellaren Instabilitätszonen. *Zeitschrift für Astrophysik, 28*, 304.

Biermann, L. (1951b). Kometenschweife und solare Korpuskularstrahlung. *Zeitschrift für Astrophysik, 29*, 271.

G. Rüdiger, *The Astrophysical Observatory Potsdam - Triumph and Tragedies*, Astronomers' Universe, https://doi.org/10.1007/978-3-032-06294-9

"""

Bigg, C. (2008). Die Himmeskarte von Potsdam aus gesehen. In: J. Lamy.

Birck, O., & Pahlen, E. v.d. (1921). Bericht über die internationale Astronomenversammlung in Potsdam. *Die Naturwissenschaften, 9*, 839.

Bleyer, U. et al. (1979). Zur Geschichte der Theorie der Lichtausbreitung. *Die Sterne 55*, Heft 1.

Boch, R. (2008). *Der Potsdamer Telegrafenberg—ein traditionsreicher Forschungsstandort zwischen DDR und wiedervereinigtem Deutschland.* GeoForschungsZentrum.

Bollé, M. (1993). Die Observatorien auf dem Telegraphenberg. *Brandenburgische Denkmalpflege, 2*, 73.

Brosche, P. (1962). Zum Masse-Drehimpuls-Diagramm von Doppel-und Einzelsternen. *Astronomische Nachrichten, 285*, 26.

Brosche, P. (2014). Der Doppel-Astronom. Erinnerungen an Ullrich Güntzel-Lingner anlässäich seines 100. Geburtstags. *Acta Historica Astronomiae, 50*, 213.

Brück, H. A. (1942a). The structure of the galaxy. *The Observatory, 64*, 326.

Brück, H. A. (1942b). Obituary notices: Friedrich Wilhelm Hans Ludendorff. *Monthly Notices of the Royal Astronomical Society, 102*, 78.

Brück, H. A. (2000). Recollection of life as a student and a young astronomer in Germany in the 1920s. *Journal of Astronomical History and Heritage, 3*, 115.

Bruggencate, P. T., & Klüber, H. V. (1947). Beobachtung magnetischer Felder auf der Sonne. *FIAT Review of German Science.*

Brunnckow, K., & Grotrian, W. (1949). Über die zeitliche Änderung der magnetischen Feldstärke von Sonnenflecken im Laufe eines Tages. *Zeitschrift für Astrophysik, 26*, 31.

Bumba, V., & Kleczek, J. (1976). *Basic mechanisms of solar activity.* D. Reidel Publishing Company.

Buthmann, R. (2020). *Versagtes Vertrauen, Wissenschaftler der DDR im Visier der Staatssicherheit.* Vandenhoeck & Ruprecht.

Carrington, R. C. (1859). On the distribution of the solar spots in latitudes since the beginning of the year 1854, with a map. *Monthly Notices of the Royal Society, 19*, 1.

Carrington, R. C. (1863). *Observations of the spots of the sun from November 9, 1853, to March 24, 1861 made at Redhill.* Williams and Norgeta.

Clerke, A. M. (1887). Richard Christopher Carrington. In: *The dictionary of national biography 1885–1900* (vol. 9).

Cliver, E. W. (2005). Carrington, Schwabe, and the gold medal. *EOS, 86*, 43.

Cliver, E. W., & Keer, N. C. (2012). Richard Christoph Carrington: Briefly among the great scientists of his time. *Solar Physics, 280*, 1.

Cowley, C. R., et al. (1986). *Upper Main sequence stars with anomalous abundances.* D. Reidel Publishing.

Deubner, F.-L., et al. (1979). Solar p-mode oscillations as tracers of radial differential rotation. *Astronomy and Astrophysics, 72*, 177.

Dick, W. R. (1988). The Potsdam zone of the astrographic catalogue. *Bulletin d'Information du Centre de Donnees Stellaires, 34*, 155.

Dick, W. R. (2000). 300 Jahre Astronomie in Berlin und Potsdam—ein Überblick. *Acta Historica Astronomiae, 8,* 11.

Dick, W. R., & Ackermann, P. (2025). Astronomie in der DDR. *Acta Historica Astronomiae, 75.* Akademische Verlagsanstalt Leipzig.

Eberhard, G., & Schwarzschild, K. (1913). The reversal of the calcium lines H and K in stellar spectra. *Astrophysical Journal, 38,* 292.

Eddy, J. A. (1975). The case of the missing sunspots. *Bulletin of the American Astronomical Society, 7, 365.*

Eddy, J. A. (1976). The Maunder Minimum. *Science, 192,* 1189.

Eggers, B. (1995). Die Geschichte des Telegrafenbergs in Potsdam bis 1900. In: *Der Einsteinturm in Potsdam.* Ars Nicolai.

Einstein, A. (2020). *Gesammelte Schriften* Bd. 8A. University Press.

Elsässer, H. (1995). Zur Schließung der Sternwarte Sonneberg. *Sterne und Weltraum, 34,* 276.

Elstner, D., et al. (1990). Galactic dynamo models without sharp boundaries. *Geophysical & Astrophysical Fluid Dynamics, 50,* 85.

Elstner, D., et al. (1992). Galactic dynamos and their radio signatures. *Astronomy and Astrophysics Supplement, 94,* 58.

Entzian, G. (2020). *Von der Rostocker Luftwarte zum Observatorium für Ionosphärenforschung Kühlungsborn.* s.n.

Finlay-Freundlich, E. (1930). Über den heutigen Stand des Nachweises der von der Relativitätstheorie behaupteten Rotverschiebung der Fraunhoferschen Linien im Sonnenspektrum. *Forschungen und Fortschritte, 61,* 8.

Finlay-Freundlich, E. (1953). Der heutige Stand der empirischen Bestätigung der allgemeinen Relativitätstheorie. *Physikalische Blätter, 9,* 14.

Finlay-Freundlich, E. (1969). Wie es dazu kam, daß ich den Einsteinturm errichtete (etwa von 1961). *Physikalische Blätter, 25,* 538.

Foerster, W. (1875). Über einige neue, mit der Berliner Sternwarte verbundene, astronomische Institutionen. *Vierteljahrsschrift der Astronomischen Gesellschaft, 10,* 268.

Foerster, W. (1911). *Lebenserinnerungen und Lebenshoffnungen Kap. 17.* G. Reimer.

Freundlich, E., et al. (1930). Über den Verlauf der Wellenlängen der Fraunhoferschen Linien längs der Sonnenoberfläche. *Zeitschrift für Astrophysik, 1,* 43.

Freundlich, E. (1931). Über die Ablenkung des Lichtes im Schwerefeld der Sonne. *Forschungen und Fortschritte, 7,* 292.

Freundlich, E., et al. (1931). Ergebnisse der Potsdamer Expedition zur Beobachtung der Sonnenfinsternis von 1929, Mai 9, in Takengon (Nordsumatra). *Zeitschrift für Astrophysik, 3,* 17.

Friedman, H., et al. (1951). Photon counter measurements of solar X-rays and extreme ultraviolet light. *Physical Review, 83,* 1025.

Fröhlich, H.-E., & Marx, S. (1990). *Astronomie in der DDR—eine Studie.* ZIAP.

Gailitis, A., et al. (2000). Detection of a flow induced magnetic field eigenmode in the Riga dynamo facility. *Physical Review Letters, 84,* 4365.

Gerth, E., et al. (1991). Magnetic field measurements of the supergiant ν Cep. *Astronomische Nachrichten, 312,* 107.

Gerth, E., et al. (1999). Magnetic field and radial velocity of the CP2 star alpha2CVn. *Astronomy and Astrophysics, 351,* 133.

Gottlöber, S., et al. (1990a). *Early evolution of the universe and formation of structure.* Akademie-Verlag Berlin.

Gottlöber, S., et al. (1990b). Sixth-order gravity and conformal transformations. *Classical and Quantum Gravity, 7,* 893.

Greiner, J., Naumann, C., & Wenzel, W. (1991). Optical outburst image inside the gamma-ray burst source 0008+13 error box. *Astronomy and Astrophysics, 242,* 425.

Gressel, O., et al. (2008). Direct simulations of a supernova-driven galactic dynamo. *Astronomy and Astrophysics, 486,* L35.

Grotrian, W. (1937). Zur physikalischen Deutung der Lichtkurve der Nova Herculis 1934. *Zeitschrift für Astrophysik, 13,* 215.

Grotrian, W. (1939). Zur Frage der Deutung der Linien im Spektrum der Sonnenkorona. *Naturwissenschaften, 27,* 214.

Grotrian, W. (1952). 30 Jahre Forschungsarbeit im Einsteinturm Potsdam. *Wissenschaftliche Annalen, 1,* 79.

Grotrian, W. (1956). Polaritäten und Maximalwerte magnetischer Feldstärken von Sonnenflecken in den Jahren 1952-1953. *Publikationen des Astrophysikalischen Observatoriums zu Potsdam 30.*

Grotrian, W., & Rambauske, W. (1935). Über das Spektrum der Nova Herculis 1934. *Zeitschrift für Astrophysik, 10,* 209.

Gruner, W. (2008). *Die Verfolgung und Ermordung der europäischen Juden durch das nationalsozialistische Deutschland, Bd. 1 1933–1937.* Oldenbourg.

Güntzel-Lingner, U. (1955). Bahnbestimmung von drei Doppelsternen mit großer Parallaxe. *Astronomische Nachrichten, 282,* 183.

Gürtler, J., & Dorschner, J. (1992). Unvergessene Politastronomie. *Die Sterne, 68,* Heft 2.

Gußmann, E. -A., & Dick, W. R. (2000). Hermann Carl Vogels Bericht über eine Reise nach England, Schottland und Irland im Jahr 1875. *Acta Historica Astronomiae, 11,* 97.

Hale, G. E. (1908). On the probable existence of a magnetic field in sunspots. *Astrophysical Journal, 28,* 315.

Hartmann, J. (1904). Investigations of the spectrum and orbit of delta Orionis. *Astrophysical Journal, 19,* 268.

Hartmann, J. (1908). Untersuchungen über das 80-cm-Objektiv des Potsdamer Refraktors. *Publikationen des Astrophysikalischen Observatoriums zu Potsdam, 15,* 2.

Hassenstein, W. (1941). Das Astrophysikalische Observatorium Potsdam in den Jahren 1875-1939. *Mitteilungen des Astrophysikalischen Observatoriums, 1.*

Helmbold, B. (2016). *Forschungstechnologien und Wissenschaftspolitik in der Biografie des Physikers Max Steenbeck (1904–1981).* Dissertation Friedrich-Schiller-Universität Jena.

Hempelmann, A., & Kurths, J. (1990). Dynamics of the outburst series of SS Cygni. *Astronomy and Astrophysics, 232*, 356.

Hentschel, K. (1992). *Der Einsteinturm.* Spektrum Akademischer Verlag.

Hentschel, K. (1995). Physik, Astronomie und Architektur—Der Einsteinturm als Resultat des Zusammenwirkens von Einstein, Freundlich und Mendelsohn. In: *Der Einsteinturm in Potsdam.* Ars Nicolai.

Hermann, A. (1996). *Einstein, der Weltweise und sein Jahrhundert.* Piper.

Herrmann, D. B. (1975). Zur Vorgeschichte des Astrophysikalischen Observatoriums Potsdam. *Astronomische Nachrichten, 296*, 245.

Herschel, W. (1801). Observations tending to investigate the nature of the sun, in order to find the causes or symptoms of its variable emission of light and heat; with remarks on the use that may possibly be drawn from solar observations. *Philosophical Transactions of the Royal Society A, 91*, 265.

Hertzsprung, E. (1911). Über die Verwendung photographischer effektiver Wellenlaengen zur Bestimmung von Farbenaequivalenten vol 22. *Publicationen des Astrophysikalischen Observatoriums zu Potsdam 3.*

Hertzsprung, E. (1917). Karl Schwarzschild. *Astrophysical Journal, 45*, 285.

Hertzsprung, E. (1920). Photographische Messungen von Doppelsternen von 1914.0 bis 1919.0. *Publicationen des Astrophysikalischen Observatoriums zu Potsdam, 24, 3.*

Hoffmann, D. (2015). In den Fussstapfen von Einstein: Der Physiker Achilles Papapetrou. in Ost-Berlin. In: *Deutsch-Griechische Beziehungen im ostdeutschen Staatssozialismus (1949–1989).*

Hoffmeister, C. (1923). Anzeige des Todes von Carl Ernst Albrecht Hartwig. *Astronomische Nachrichten, 219*, 185.

Hoffmeister, C. (1929). Die Entdeckung neuer veränderlicher Sterne. *Die Sterne, 9,* 1.

Hoffmeister, C. (1957). Über das photometrische Verhalten einiger RW Aurigae-Sterne. *Communications of the Konkoly Observatory, 42*, 13.

Hoffmeister, C. (1962). Über eine intergalaktische Absorptionswolke. *Zeitschrift für Astrophysik, 55,* 46.

Hoffmeister, C. (1965). Colloquium über Veränderliche Sterne in Bamberg 1965. *Die Sterne, 41,* 197.

Hoffmeister, C. (1968). Mitteilung über neuentdeckte Veränderliche Sterne. *Astronomische Nachrichten, 290*, 276.

Hoppe, J. (1962). Prof. Dr. Cuno Hoffmeister zum 70. Geburtstag. *Die Sterne, 38,* 120.

Howard, R., & LaBonte, B. J. (1980). The sun is observed to be a torsional oscillator with a period of 11 years. *Astrophysical Journal, 239*, L33.

Hubrig, S., & Kurths, J. (1989). Sternbilder mit dem Doppler-Imaging-Verfahren. *Die Sterne, 65,* Heft 6.

Humboldt, A. V. (2004). *Kosmos, Entwurf einer physischen Weltbeschreibung, III,* 543. Eichborn.

Jäger, F. W. (1972). Aspekte der solar-terrestrischen Forschung. *Die Sterne, 48,* 87.

Jäger, F. W. (1979). Astronomische Grundlagen zur technischen Nutzung der Sonnenenergie. *Die Sterne, 55,* 144.

Kaisig, M., et al. (1993). The alpha-effect due to supernova explosions. *Astronomy and Astrophysics, 274,* 757.

Keinigs, R. K. (1983). A new interpretation of the alpha effect. *Physics of Fluids, 26,* 2558.

Kelch, W. L., et al. (1978). Stellar model chromospheres VII. Capella, Pollux and Aldebaran. *Astrophysical Journal, 222,* 931.

Kempf, P. (1895). Meteorologische Beobachtungen in den Jahren 1888 bis 1893. *Publikationen des Astrophysikalischen Observatoriums zu Potsdam, 10,* 2.

Kempf, P. (1905). The spectroheliograph of the Potsdam observatory. *Astrophysical Journal, 21,* 49.

Kempf, P. (1916). Bestimmung der Rotation der Sonne aus der Bewegung von Kalziumflocken. *Publikationen des Astrophysikalischen Observatoriums zu Potsdam, 23,* 1.

Kienle, H. (1949). *Ein 2 m-Universal-Spiegelteleskop.* In: Miscellanea Academica Berolinensia, Gesammelte Abhandlungen, 25.

Kienle, H. (1956). Nachruf auf Walter Grotrian. *Mitteilungen der Astronomischen Gesellschaft, 7,* 5.

Kippenhahn, R. (1963). Differential rotation in stars with convective envelopes. *Astrophysical Journal, 137,* 664.

Kippenhahn, R. (1965). Stellar evolution and variability. *Kleine Veröffentlichungen der Remeis-Sternwarte Bamberg Bd. IV, Nr. 40,* 7.

Kippenhahn, R. (1968). Nachruf auf Cuno Hoffmeister. *Mitteilungen der Astronómischen Gesellschaft, 24,* 5.

Kippenhahn, R. (1999). Operativ-Vorgang Horoskop. *Star-Observer, 12,* 68.

Kippenhahn, R. (2001). Liebesgrüße aus Prag. *Sterne und Weltraum, 3,* 226.

Kirsten, C., & Treder, H.-J. (1979). *Albert Einstein in Berlin.* Akademie-Verlag Berlin

Kitchatinov, L. L., & Rüdiger, G. (1999). Differential rotation models for late-type dwarfs and giants. *Astronomy and Astrophysics, 344,* 911.

Klein, J. (1869). Die Ergebnisse der Beobachtungen der totalen Sonnenfinsternis vom !8. August 1868. Gaea 5.

Klüber, H. v. (1947). Ionosphäre und Sonnenforschung. *Funk und Ton 2.*

Klüber, H. v. (1965). Erwin Finley-Freundlich (Nachruf). *Astronomische Nachrichten, 288,* 281.

Knölker, M., & Stix, M. (1983). A convenient method to obtain stellar eigenfrequencies. *Solar Physics, 82,* 334.

Krause, F., & Beck, R. (1998). Symmetry and direction of seed magnetic fields in galaxies. *Astronomy and Astrophysics, 335,* 789.

Krause, F., & Rädler, K.-H. (1980). *Mean-field magnetohydrodynamics and dynamo theory.* Pergamon Press.

Krause, M., et al. (1989). The magnetic field structures in two nearby spiral galaxies. II. The bisymmetric spiral magnetic field in M81. *Astronomy & Astrophysics, 217,* 17.

Krause, F., et al. (1990). Key problems of flat objects dynamo theory and ways of their solution. In: R. Beck et al. (ed.), *Galactic and intergalactic magnetic fields*. Kluwer Academic Publishers.

Krug, W. (1937). Photometrische Bearbeitung der galaktischen Sternhaufen M71 und Harv. 20. *Zeitschrift für Astrophysik, 13*, 205.

Kühn, G. (2012). Chronik des Miethe-Teleskops auf dem Chemie-Gebäude der TH Charlottenburg. In: A. Miethe (Eds.), *Lebenserinnerungen (Acta Historica Astronomiae 46)*. Frankfurt am Main: Verlag Harri Deutsch.

Kuiper, G. P. (1946). German astronomy during the war. *Popular Astronomy, 54*, 263.

Kukutz, I. (2009). *Die Gründung des Neuen Forums*. Robert-Havemann-Gesellschaft.

Kummer, H.-J. (1996). Hans Kienle. Ein Lebensbild zu seinem 100. Geburtstag. *Sterne u. Weltraum, 4*, 266.

Kurths, J. (1989). Mathematische Methoden der Datenanalyse—ein unvermeidliches Hilfsmittel? *Die Sterne, 65*, 243.

Lamy, J. (2008). *La Carte du Ciel*. EDP Sciences.

Lauter, E. A. (1975a). Entwicklungstendenzen der solar-terrestrischen Physik. *Sitzungsberichte der Akademie der Wissenschaften der DDR/Mathematik, Naturwissenschaften, Technik* 35.

Lauter, E. A. (1975b). Bedeutung der Erforschung des erdnahen Raumes für den Menschen. In:. *Weltraum und Erde*. Transpress Verlag.

Lauter, E. A. (1984). Einige Bemerkungen zu gegenwärtigen Trends in den solar-terrestrischen Beziehungen. *Aus der Arbeit von Plenum und Klassen der AdW der DDR* 9.

Lohse, O. (1889). Beschreibung des Heliographen. *Publikationen des Astrophysikalischen Observatoriums zu Potsdam, 19*, 473.

Lohse, O. (1895). Gustav Spörer (Nachruf). *Vierteljahrsschrift der Astronomischen Gesellschaft, 30*, 208.

Lohse, O. (1907). Hermann Carl Vogel (Nachruf). *Astronomische Nachrichten, 175*, 373.

Ludendorff, H. (1932). Über die Ablenkung des Lichtes im Schwerefeld der Sonne. *Astronomische Nachrichten, 244*, 321.

Mattig, W. (1999). Walter Grotrians fundamentale Beiträge zur Physik der Sonnenkorona. *Sterne u. Weltraum, 6–7*, 557.

Maunder, E. W. (1890). Prof. Spoerer's researches on sunspots. *Monthly Notices of the Royal Society, 50*, 251.

Maunder, E. W. (1894). A prolonged sunspot minimum. *Knowledge: An illustrated magazine of science, 17*, 173.

Meinel, R. (1994). Karrieremuster. *Spektrum der Wissenschaft*, Heft 5.

Meinel, R., & Brandenburg, A. (1990). Behaviour of highly supercritical alpha-effect dynamos. *Astronomy and Astrophysics, 238*, 369.

Meinhold, G. (2014). *Der besondere Fall Jena—Die Universität im Umbruch 1989–1991*. Franz Steiner Verlag.

Meinunger, L. (1979). Discovery of a period in the symbiotic star AG Draconis. *Information Bulletin on Variable Stars, 1611*, 2.

Mestel, L. (1984). Theory of magnetic stars. *Astronomische Nachrichten, 305*, 301.

Miethe, A. (2012). Lebenserinnerungen. *Acta Historica Astronomiae 46.*

Miethe, A., et al. (1916). *Die totale Sonnenfinsternis vom 21. August 1914 beobachtet in Sandnessjöen auf Alsten (Norwegen).* Fr. Vieweg & Sohn.

Milne, E. A. (1931). Dense stars. *The Observatory, 54*, 140.

Moffatt, H. K. (1973). An approach to a dynamic theory of dynamo action in a rotating conducting fluid. *Journal of Fluid Mechanics, 53*, 385.

Moffatt, H. K. (1978). *Magnetic field generation in electrically conducting fluids.* University Press.

Möhlmann, D. (1984). Azimuthal instability and radial structure of the pre-planetary disk. *Gerlands Beiträge zur Geophysik, 93*, 231.

Müller, G. (1907). Hermann Carl Vogel (Nachruf). *Vierteljahrsschrift der Naturforschenden Gesellschaft, 42*, 323.

Müürsepp, P. (1982). *Die Jugendjahre von Bernhard Schmidt und sein Briefwechsel mit dem Potsdamer Observatorium.* Valgus Tallin.

Oetken, L. (1966). Magnetische Sterne. *Mitteilungen der Astronomischen Gesellschaft, 21*, 27.

Oetken, L., et al. (1970). Spektroskopische Untersuchungen des magnetischen Sterns alpha2CVn. *Astronomische Nachrichten, 292*, 1.

Pahlen, E. v.d. (1937). *Lehrbuch der Stellarstatistik. Unter Mitwirkung von F. Gondolatsch.* Barth.

Pais, A. (1986). *Raffiniert ist der Herrgott… Albert Einstein, eine wissenschaftliche Biographie.* Vieweg.

Pedde, B. (2023). Orientalismus auf dem Telegrafenberg in Potsdam. *Mitteilungen, Informationen, Programm. Wilhelm-Förster-Sternwarte, 18*, 12.

Pfau, W., & Schielicke, R. E. (2013). Wir sind wohl doch ein bißchen zu sehr in Illusionen gewesen—Die politische Geschichte der Astronomischen Gesellschaft im geteilten Deutschland. In: D. Lemke (ed.), *Die Astronomische Gesellschaft 1863–2013. Bilder und Geschichten aus 150 Jahren.* Astronomische Gesellschaft.

Piskunov, N. E., et al. (1990). Surface imaging of late-type stars. *Astronomy and Astrophysics, 230*, 363.

Rädler, K.-H., & Bräuer, H.-J. (1987). On the oscillatory behaviour of kinematic mean-field dynamos. *Astronomische Nachrichten, 308*, 101.

Reinhold, T., et al. (2013). Rotation and differential rotation of active Kepler stars. *Astronomy and Astrophysics, 560*, 4.

Richter, G. A. (1992). Zur Geschichte der Sternwarte Sonneberg. In S. Marx (Ed.), *Cuno Hoffmeister. Festschrift zum 100. Geburtstag.* J.A. Barth Verlag.

Richter, M. (2015). *Nikolaus Benjamin Richter, Astronom, Meteorologe, Geograph, Maler und Vater.* s.n.

Richter, G. A., & Scholz, R.-D. (1990). Absolute proper motions of blue objects in the vicinity of M33. *Astronomische Nachrichten, 311*, 31.

Richter, G., & Wenzel, W. (1984). *Veränderliche Sterne*. Johann Ambrosius Barth.

Roberts, P. H., & Stix, M. (1971). *The turbulent dynamo*. A translation of papers by F. Krause, K.-H. Rädler and M. Steenbeck. NCAR Technical Notes.

Ruben, G. (1991). Cosmic rotation and the inertial system. *Ap&SS, 177*, 465.

Rüdiger, G. (1980). Rapidly rotating α2 dynamo models. *Astronomische Nachrichten, 301*, 181.

Rüdiger, G. (1984). Sunspot story. *Die Sterne, 60*, Heft 5.

Rüdiger, G. (1989). *Differential rotation and stellar convection: Sun and solar-type stars*. Gordon and Breach Science Publishers/Akademie-Verlag Berlin.

Rüdiger, G., & Kitchatinov, L.L. (1990). The turbulent stresses in the theory of the solar torsional oscillations. *Astronomy and Astrophysics, 236*, 503.

Rüdiger, G. (2024). Astronomen, Akten und Affären. *Acta Historica Astronomiae, 71*. Akademische Verlagsanstalt Leipzig

Rüdiger, G., & Kitchatinov, L. L. (1997). The slender solar tachocline: A magnetic model. *Astronomische Nachrichten, 318*, 273.

Runge, C. (1916). Karl Schwarzschild. *Physikalische Zeitschrift, 17*, 545.

Ruzmaikin, A. A. et al. (1988). *Magnetic fields of galaxies*. Kluwer.

Saal, E. (1901). *Das Kuppelgebäude für den Grossen Refraktor des Astrophysikalischen Observatoriums auf dem Telegraphenberge bei Potsdam*. W. Ernst & Sohn.

Scheiner, J. (1890). *Die königlichen Observatorien für Astrophysik, Meteorologie und Geodäsie bei Potsdam, aus amtlichem Anlass herausgegeben von den beteiligten Direktoren*. Mayer & Müller.

Scheiner, J. (1897). *Die Photographie der Gestirne*. W. Engelmann.

Schellbach, K. (1890). *Erinnerungen an unseren Kronprinzen Friedrich Wilhelm von Preußen*. Eduard Trewendt.

Schielicke, R. E. (2013). *Wer kennt die Völker—nennt die Namen…Die Astronomische Gesellschaft und ihre Mitglieder 1864–2013*. Astronomische Gesellschaft.

Schnell, G. (2005). *Das Lindenhotel, Berichte aus dem Potsdamer Geheimdienstgefängnis*. Ch. Links Verlag.

Scholz, G. (1971). High effective magnetic field strengths of the magnetic star 53 Camelopardalis. *Astronomische Nachrichten, 292*, 281.

Scholz, G. (2000). Über einige wissenschaftliche Beiträge aus den ersten Jahrzehnten des Astrophysikalischen Observatoriums Potsdam. *Acta Historica Astronomiae, 11*, 97.

Scholz, G., & Gerth, E. (1980). Radial velocity and magnetic field measurements of the A-type supergiant v Cep (HD 207260). *Astronomische Nachrichten, 301*, 211.

Schorr, R. (1936). Bernhard Schmidt (Nachruf). *Astronomische Nachrichten, 258*, 455.

Schröter, E. H. (1955). Zur Deutung der Rotverschiebung im Sonnenspektrum. *Die Naturwissenschaften, 42*, 623.

Schröter, E. H. (1957). Zur Deutung der Rotverschiebung und der Mitte-Rand-Variation der Fraunhoferlinien bei Berücksichtigung der Temperaturschwankungen der Sonnenatmosphäre. *Zeitschrift für Astrophysik, 41*, 141.

Schubart, J. (1961). Definitive Elemente des Kometen 1959d (Bester-Hoffmeister). *Astronomische Nachrichten, 286*, 7.

Schubart, J. (1962). Eindrücke von der IAU Tagung in Californien. *Die Sterne, 38*, 118.

Schwabe, S. H. (1844). Sonnen-Beobachtungen im Jahre 1843. *Astronomische Nachrichten, 21*, 233.

Schwarzschild, K. (1906). Ueber das Gleichgewicht der Sonnenatmosphäre. *Nachrichten von der Gesellschaft der Wissenschaften zu Göttingen, 1*, 41.

Schwarzschild, K. (1914). Über die Verschiebung der Bande bei 3883 A im Sonnenspektrum. *Sitzungsberichte der Königlich Preußischen Akademie der Wissenschaften, 1201*.

Seegert, B. (1927). Adolf Miethe (Nachruf). *Astronomische Nachrichten, 230*, 205.

Seehafer, N. (1990). Electric current helicity in the solar atmosphere. *Solar Physics, 125*, 219.

Seiler, M. P. (2007). Kommandosache Sonnengott, Geschichte der deutschen Sonnenforschung im Dritten Reich. *Acta Historica Astronomiae, 31*. Verlag Harry Deutsch.

Shankland, R. S. (1982). Michelson in Potsdam. *Astronomische Nachrichten, 303*, 3.

Solanki, S. K., & Fligge, M. (1998). Solar irradiance since 1874 revisited. *Geophysical Research Letters, 25*, 341.

Sommeria, J., & Moreau, R. (1982). Why, how, and when MHD turbulence becomes two-dimensional. *Journal of Fluid Mechanics, 118*, 507.

Soward, A. M. (2023). Paul Harry Roberts. 13 September 1929–17 November 2022. In: *Biographical Memoirs of Fellows of the Royal Society*.

Spiegel, A. (2002). *Die Stasi kam im Morgengrauen. Jugendlicher Widerstand in Werder (Havel) 1950 bis 1953.* Heimat- und Fremdenverkehrsverein Werder und Stadt Werder.

Spieker, P. (1879). *Baubericht über die technischen Anlagen für das Königliche Astrophysikalische Observatorium auf dem Telegraphenberg bei Potsdam.* Ernst & Korn.

Spieker, P. (1894). Die Königlichen Observatorien für Astrophysik, Meteorologie und Geodäsie auf dem Telegraphenberge bei Potsdam. *Zeitschrift für Bauwesen, 46*, 1.

Spörer, G. (1861). Beobachtungen von Sonnenflecken und daraus abgeleitete Elemente der Rotation der Sonne. *Astronomische Nachrichten, 55*, 289.

Spörer, G. (1869). *Die Reise nach Indien.* Vortrag in der Singakademie.

Spörer, G. (1871). Über die Vergleichung von Sonnenflecken und Protuberanzen. *Astronomische Nachrichten, 78*, 81.

Spörer, G. (1873). Beobachtungen von Sonnenflecken und Protuberanzen. *Astronomische Nachrichten, 82*, 138.

Spörer, G. (1874). *Beobachtungen der Sonnenflecken zu Anclam.* Leipzig.

Spörer, G. (1887). Über die Periodicität der Sonnenflecken seit dem Jahre 1618 .... und Nachweis einer erheblichen Störung dieser Periodicität während eines langen Zeitraumes. *Vierteljahrsschrift der Astronomischen Gesellschaft, 22*, 323.

Spörer, G. (1894). Beobachtung von Sonnenflecken in den Jahren 1885 bis 1893. *Publicationen des Astrophysikalischen Observatoriums zu Potsdam, 10*, 1. Stück.

St. John, C. E. (1910). The general circulation of the mean and high-level calcium vapor in the solar atmosphere. *Astrophysical Journal, 32*, 36.

Stark, I. (1997). Der Runde Tisch der Akademie und die Reform der AdW nach der Herbstrevolution 1989: Ein gescheiterter Versuch der Selbsterneuerung. *Geschichte und Gesellschaft, 23*.

Steenbeck, M. (1961). Skizze zu einer magnetohydrodynamischen Theorie der Sonnenfleckenvorgänge. *Mitteilungen a. d. Institut für Magnetohydrodynamik der AdW zu Berlin, Jena, S. 153*.

Steenbeck, M., & Krause, F. (1965). Elektromagnetische Rückkopplung durch Turbulenz unter Einwirkung von Corioliskräften. *Mitteilungen der Astronomischen Gesellschaft, 19*.

Steenbeck, M., et al. (1966). Berechnung der mittleren Lorentz-Feldstärke für ein elektrisch leitendes Medium in turbulenter, durch Coriolis-Kräfte beeinflußter Bewegung. *Zeitschrift für Naturforschung, 21*, 1285.

Steenbeck, M., et al. (1968). Experimental discovery of the electromotive force along the external magnetic field induced by a flow of liquid metal (alpha-effect). *Soviet Physics Doklady, 13*, 443.

Stępień, K. (1984). Photometry and spectroscopy of magnetic stars. *Astronomische Nachrichten, 305*, 311.

Stix, M. (1975). The galactic dynamo. *Astronomy and Astrophysics, 42*, 85.

Stix, M. (1984). Solar magnetism: Observations and theory. *Astronomische Nachrichten, 305*, 215.

Strassmeier, K., et al. (2004). The STELLA robotic observatory. *Astronomische Nachrichten, 325*, 527.

Tiemann, K. -H. (1991). Hans Ludendorff zum 50. Todestag. *Academie Spectrum, 22*, Heft 7.

Tuominen, I., et al. (1991). *The sun and cool stars: Activity, magnetism, dynamos*. Springer-Verlag.

Vogel, H. C. (1872/1873). *Beobachtungen angestellt auf der Sternwarte des Kammerherrn v. Bülow zu Bothkamp Hefte 1/2*. W. Engelmann.

Vogel, H. C. (1890a). Spectrographische Beobachtungen an Algol. *Astronomische Nachrichten, 123*, 289.

Vogel, H. C. (1890b). Spektrographische Entdeckung einer Bahnbewegung des Sterns α Virginis. *Naturwissenschaftliche Rundschau, 25*, 313.

Vogel, H. C. (1892). List of the proper motions in the line of sight of fifty-one stars. *Monthly Notices of the Royal Astronomical Society, 52*, 541.

Vogel, H. C. (1895). Todesanzeige Friedrich Wilhelm Gustav Spörer. *Astronomische Nachrichten, 138*, 247.

Vogel, H. C. (1904). Jahresbericht Potsdam. *Vierteljahrsschrift der Astronomischen Gesellschaft, 39*, 122.

Vogel, H. C. (1907). Die zwei Doppelrefraktoren des Observatoriums. *Publicationen des Astrophysikalischen Observatoriums zu Potsdam, 45*, 1.

Waldmeier, M. (1941). *Ergebnisse und Probleme der Sonnenforschung*. Becker & Erler.

Ward, F. (1965). The general circulation oft the solar atmosphere and the maintenance of the equatorial acceleration. *Astrophysical Journal, 141*, 534.

Weber, T. (2006). *Dr. Paul Ahnert (1897-1989), Leben und Werk eines Astronomen im Spannungsfeld von Wissenschaft, Popularisierung und Politik*. FernUniversität Hagen, Magisterarbeit.

Wempe, J. (1955). Walter Grotrian (Nachruf). *Astronomische Nachrichten, 228*, 190.

Wempe, J. (1975). Zum 100. Jahrestag der Gründung des Astrophysikalischen Observatoriums Potsdam. *Die Sterne, 51*, 193.

Wempe, J. (1976). Hans Kienle (22.10.1895–15.2.1975 Nachruf). *Astronomische Nachrichten, 297*, 99.

Wilsing, J. (1888). Ableitung der Rotationsbewegung der Sonne aus Positionsbestimmungen von Fackeln. *Astronomische Nachrichten, 119*, 311.

Wilsing, J. (1914). Julius Scheiner (Nachruf). *Vierteljahrsschrift der Astronomischen Gesellschaft, 49*, 22.

Wolf, R. (1861). *Die Sonne und ihre Flecken*. Orell & Co. Zürich.

Wolf, R. (1877). *Geschichte der Astronomie*. R. Oldenbourg.

Index of Persons and Firms

GPSR Compliance
The European Union's (EU) General Product Safety Regulation (GPSR) is a set
of rules that requires consumer products to be safe and our obligations to
ensure this.

If you have any concerns about our products, you can contact us on

ProductSafety@springernature.com

In case Publisher is established outside the EU, the EU authorized
representative is:

Springer Nature Customer Service Center GmbH
Europaplatz 3
69115 Heidelberg, Germany